BRITISH NEARSHORE FORAMINIFERIDS

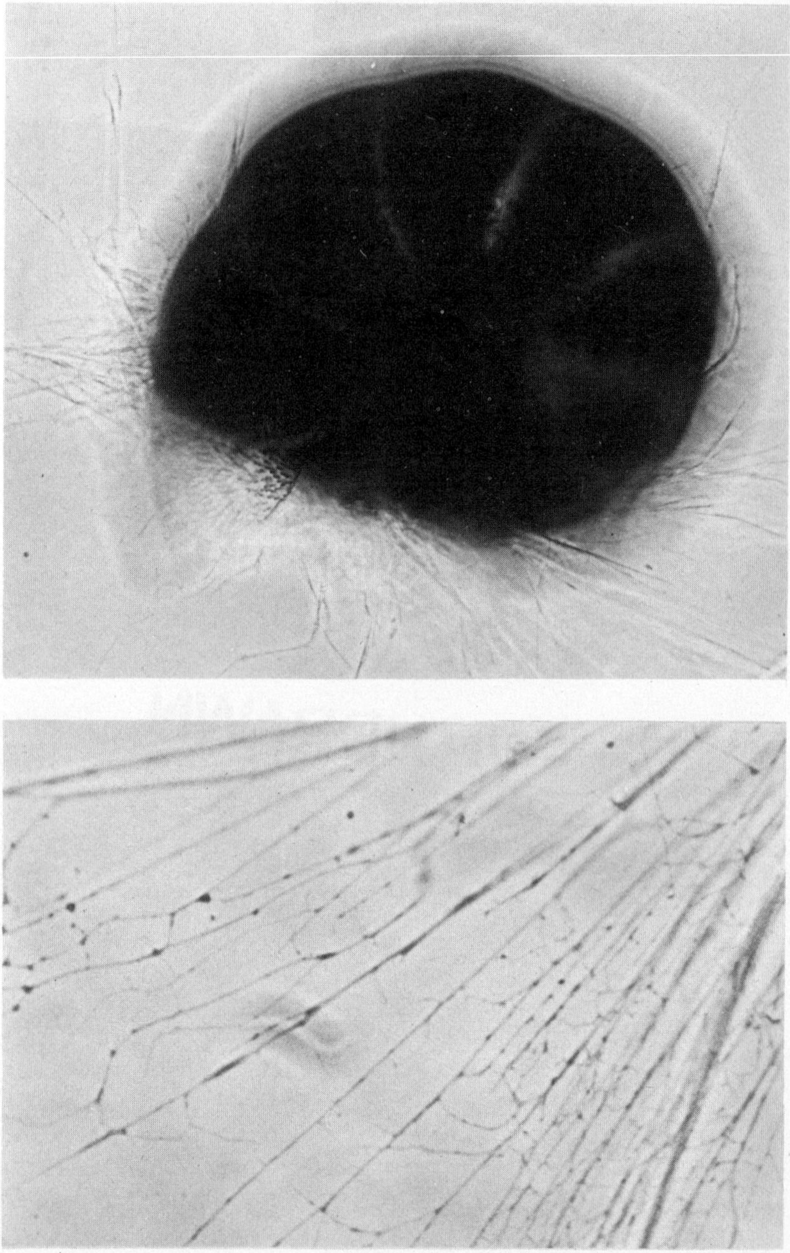

Frontispiece. Protelphidium germanicum (Ehrenberg) from the Exe estuary. *Top:* general view to show pseudopodia ×250. *Bottom:* detail of the pseudopodial net and granules ×1000 approx.

A NEW SERIES

Synopses of the British Fauna

No. 16

Edited by Doris M. Kermack and R. S. K. Barnes

BRITISH NEARSHORE FORAMINIFERIDS

Keys and Notes for the Identification of the Species

JOHN W. MURRAY

Department of Geology,
University of Exeter,
Exeter, Devon, England

1979

Published for
THE LINNEAN SOCIETY OF LONDON
and
THE ESTUARINE AND BRACKISH-WATER
SCIENCES ASSOCIATION

by
ACADEMIC PRESS
LONDON, NEW YORK AND SAN FRANCISCO

ACADEMIC PRESS INC. (LONDON) LTD
24–28 Oval Road
London, NW1 7DX

U.S. Edition published by
ACADEMIC PRESS INC.
111 Fifth Avenue
New York, New York 10003

Library of Congress Catalog Card Number: 78–75267
ISBN: 0–12–511850–3

Text set in 9/10 pt VIP Times, printed by photolithography and bound in Great Britain at The Pitman Press, Bath

Foreword

British Nearshore Foraminiferids is the first synopsis devoted to a group within the subkingdom Protozoa. These small animals have a long geological history reaching back to the beginning of the Palaeozoic and possibly earlier. During this time their tests (or shells) have accumulated as bottom deposits, which have later become rock strata. Today the tests of dead foraminiferids are a major component of deep sea oozes and tropical beach sands. Thus, though small, these Protozoa are of great importance as stratigraphic markers of past climates and environments, as well as being indicators for certain commercially important deposits, such as gas and petroleum.

Because of their importance geologically, living foraminiferids have been studied by geologists rather than zoologists and thus it is appropriate that this *Synopsis* has been prepared by a professor of Geology. It has a novel approach, which is due partly to the foraminiferids being members of the subkingdom Protozoa with a completely characteristic body-form, and partly to the author, Professor John Murray, describing them from the viewpoint of a geologist. He has shown that these abundant and remarkably beautiful little creatures can be studied using quite simple equipment and that there is no longer an excuse for them to remain relatively unknown inside sixth form and undergraduate laboratories. We are all grateful to him.

R. S. K. Barnes
Estuarine and Brackish-water
Sciences Association

Doris M. Kermack
The Linnean Society

A Synopsis of the British Nearshore Foraminiferids

JOHN W. MURRAY

Department of Geology, University of Exeter,
Exeter, Devon, England

CONTENTS

Introduction

Although foraminiferids are very small members of the marine and brackish water fauna they are often present in large numbers and constitute an important element of the meiofauna. They possess hard parts in the form of tests (or shells) which on death are preserved in the sediment and are therefore of interest to geologists. The recognition of past brackish water environments is readily accomplished by a study of fossil foraminiferid assemblages.

The synopsis is aimed at those who have no prior knowledge of the group but who would like to be able to identify genera and species. For the first time a key to British brackish and nearshore genera has been prepared. Use of the key is dependent upon having mastered the morphological details described on pp. 2, 3, 12–15. It must be emphasized that foraminiferids can only readily be identified in the dry state. Their appearance changes markedly when they are immersed in water or mounted in mounting media. Sixty-three species are considered in this synopsis.

General Features

Foraminiferids are classified as protozoans because they consist of a single cell which is made up of **cytoplasm** with one or more **nuclei.** In most foraminiferids little can be seen of their soft parts because they are largely enclosed by the **test.**

The most obvious soft parts are the **pseudopodia**, protoplasmic filaments which extend from the apertural region of the test for distances of two to three times the diameter of the latter (Fig. 1; Frontispiece). The diameter of a pseudopod varies but is not generally less than 1 μm. Pseudopodia have the facility to fuse on contact so that an "active" foraminiferid will be surrounded by a net of pseudopodia. Protoplasmic streaming within the pseudopodia can be seen from the movement of lumpy granules, 2–4 μm in diameter, along the length of individual filaments. The mechanism of streaming is not understood and is not simple as granules may move in opposite directions in different parts of the same pseudopod. These features are best seen by placing a live individual in a cavity slide with its own ambient water, covering with a cover glass and leaving under a microscope (with the lamp turned off) for half to one hour. The individual may develop its **pseudopodial net** which can then be examined either with transmitted light or dark field illumination. Excessive illumination, with its consequent heating, will at best inhibit the foraminiferid or at worst kill it.

The cytoplasmic mass within the test is often coloured in benthic species. The colouration may be due to two possible causes, food and symbionts. Benthic forms living in the photic zone commonly feed on living diatoms and other algae, and on detritus. All of these are coloured, green, red or brown so the foraminiferid may also be coloured. Alternatively or in addition, symbionts may be present, especially zooxanthellae.

The granular and reticulose pseudopodia are characteristic of the subclass Granuloreticulosia of which the Foraminiferida form an order. Otherwise none of the features of the soft parts is used in taxonomy. It is the morphological features of the test which are important in this respect and these are described later (pp. 12–15).

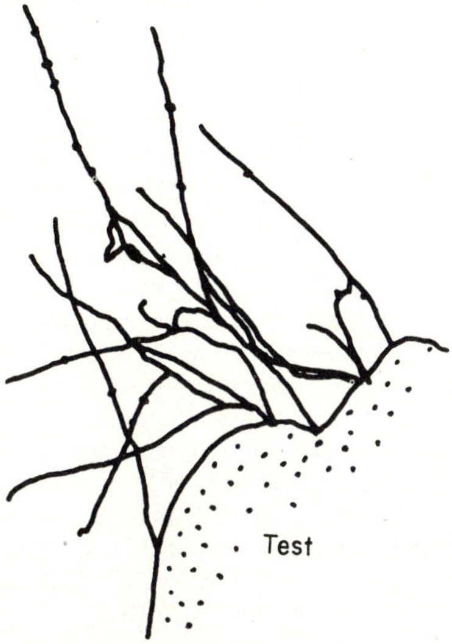

FIG. 1. Pseudopodial net emerging from the test.

Biology

Life History

The life history of very few species has been determined and even in some of these cases the precise details are not known. Sexual and asexual reproduction are known to occur and result in diploid **agamont** and haploid **gamont** generations (a characteristic of plants rather than animals).

In *Elphidium crispum* meiosis during asexual reproduction leads to the development of tiny gamonts within the parent test. These are liberated into the water and grow to become adults which then undergo mitosis to produce large numbers of flagellate gametes (2–5 μm). After fertilization in the sea an agamont is produced which grows to maturity and the cycle starts again.

A number of variations on this basic theme have been described. There is not necessarily a strict alternation from sexual to asexual generations. Sometimes there are several successive agamont generations between gamont phases. The period between reproductive phases varies from a matter of days in very favourable environments to a year or more in less favourable situations.

Movement

The means of locomotion are the pseudopodia. An active foraminiferid normally has its test oriented with the aperture facing the substrate over which the pseudopodial net spreads. During locomotion the pseudopodia show an alignment in the direction of movement. Pseudopodial streaming to a forward position takes place, then the test is pulled forward an increment, the trailing pseudopodia are shortened and the animal is ready to repeat the cycle.

Normally movement is random but directed movement takes place when an individual is escaping from an unfavourable situation, e.g. burial through an influx of sediment.

Feeding

Foraminiferids also feed by means of the pseudopodia. The latter may pick up food from the substrate or they may form an upstanding net which filters food from suspension in the water. In either event the food is moved by pseudopodial streaming to the apertural region of the test. Some genera, e.g. *Quinqueloculina*, have a mass of food around the aperture. Others, such as *Elphidium*, make a cyst-like covering of food over the whole test. Digestion is in part carried on outside the test. Forms such as *Elphidium* have apertural pores too tiny to allow food particles to enter the test.

In shallow water, microscopic algae (including diatoms) form a large part of the food. In deeper water organic detritus and bacteria are more important. Experimental evidence suggests that food is selected largely on size or specific gravity and inevitably many inorganic particles are accidentally collected too. Some species are said to be carnivorous on other foraminiferids, small sea urchins, cumaceans, etc.

The availability of bacteria as part of the diet of some species seems to be essential to healthy growth and successful reproduction.

Growth

Tubular single-chambered tests could grow continuously. However, multi-chambered tests are the product of discontinuous growth, each growth event being represented by a chamber. Thus, there are no growth lines on foraminiferid chambers.

The secretion of a chamber has rarely been observed but it appears that the pseudopodia play an important part in the shaping and laying down of the template around which the new chamber forms. The process takes a matter of hours or even days.

The patterns of growth are described on p. 14. Clearly they must be genetically controlled. However, in very variable environments such as those found in estuaries, environmentally induced abnormalities are not uncommon. These range from reduced chamber size (caused by shortage of food) to changes in coiling direction.

Enemies

Foraminiferids must serve as food for other animals and they are known to be preferentially selected by some shrimps and scaphopod molluscs. Many must be eaten by deposit-feeding invertebrates. Two instances are known where one foraminiferid species is parasitic on another.

Ecology

Although foraminiferids are known to live on the sea floor (benthic) and in suspension in the water column (planktonic), it is only the former which live in brackish and nearshore areas. Planktonic species have environmental requirements which are best met in oceanic areas.

The benthic foraminiferids of shallow water areas live on most types of substrate. Sediment-dwelling forms live both on the surface and within the upper few centimetres of the sediment, the depth varying according to the thickness of the oxidized zone but not normally deeper than 6–7 cm. Other forms prefer firm substrates such as rocks, shells or plants. Some of these forms use their pseudopodia to cling to the substrate and they can move about when necessary. Others cement themselves to the substrate and are therefore permanently fixed.

In general foraminiferids live mainly in the subtidal areas or in pools in the intertidal zone (e.g. marsh pools). They are sometimes present in intertidal muddy sands in the lower part of the intertidal zone. Dead tests, empty of protoplasm, are widely distributed from subtidal to intertidal areas. They may be locally concentrated around high water mark and instances are known where they have been blown from the shore on to adjacent sand dunes.

The main limiting factor in brackish water areas is salinity. It is the size and frequency of the variation which is important. For example, a tidal channel may be marine at high water and six hours later at low water the salinity may have dropped to 0. Thus twice daily the range would be 0–35‰. This is clearly a much more extreme environment than, say, a brackish pool in a tidal flat where the salinity range may be 20–25‰ over a much longer period.

Exposure to the atmosphere limits the distribution of the intertidal species. Substrate type also causes local differences in species distributions. Temperature is important on a regional scale but is not responsible for changes in distribution within the British Isles.

Brackish tidal marshes are characterized by:

Miliammina fusca
Jadammina macrescens
Trochammina inflata
Elphidium williamsoni

Additional species which may be present are:

Haplophragmoides wilberti
Ammobaculites balkwilli

Brackish lagoons and estuaries are characterized by:

Miliammina fusca
Ammonia beccarii
Elphidium magellanicum
Elphidium oceanensis
Elphidium williamsoni
Protelphidium germanicum

Estuary mouths and nearshore areas of near normal to normal salinity have assemblages of inner shelf species:

Reophax moniliformis
Reophax scottii
Cribrostomoides jeffreysii
Trochammina ochracea
Eggerelloides scaber
Clavulina obscura
Cyclogyra involvens
Cornuloculina balkwilli
Spiroloculina excavata
Spiroloculina rotunda
Quinqueloculina bicornis
Quinqueloculina bicornis angulata
Quinqueloculina cliarensis
Quinqueloculina dimidiata
Quinqueloculina lata
Quinqueloculina oblonga
Quinqueloculina seminulum
Massilina secans
Pateoris hauerinoides
Miliolinella subrotunda
Lagena spp.
Oolina spp.
Fissurina lucida
Fissurina marginata
Buliminella elegantissima
Bolivina pseudoplicata
Brizalina pseudopunctata
Brizalina variabilis
Bulimina marginata
Trifarina angulosa
Buccella frigida
Gavelinopsis praegeri
Rosalina anomala
Rosalina williamsoni
Glabratella millettii
Asterigerinata mamilla
Spirillina vivipara
Patellina corrugata
Ammonia beccarii
Elphidium crispum
Elphidium cuvillieri
Elphidium earlandi
Elphidium excavatum
Elphidium gerthi
Elphidium incertum
Eliphidium macellum
Elphidium margaritaceum
Elphidium oceanensis
Cibicides lobatulus
Planorbulina mediterranensis
Acervulina inhaerens
Fursenkoina fusiformis
Cassidulina obtusa
Nonion depressulus
Lamarckina haliotidea

Small species and juveniles of large species may be transported from the shelf into coastal waters during the tidal exchange of water. Such forms are recognizable by their small size (peak ~140 μm). This process operates especially in estuaries having a well mixed, lateral salinity gradient and high tidal range (e.g. Severn estuary).

World Distribution

The brackish water foraminiferids of Britain are the same as those found along most of the European seaboard (see Murray, 1971).

Brackish waters throughout the world are dominated by species of *Miliammina, Elphidium* and *Ammonia* with *Jadammina* and *Trochammina* on marshes. All these faunas are of low diversity (Murray, 1973).

The nearshore microfauna is the same as that of most of the European seaboard but certain species are known to be of southern and others of northern derivation (Murray, 1971).

Southern

Ammonia beccarii
Asterigerinata mamilla
Bolivina pseudoplicata
Elphidium crispum
Elphidium cuvillieri
Elphidium excavatum
Elphidium oceanensis
Gavelinopsis praegeri
Massilina secans
Nonion depressulus
Quinqueloculina bicornis
Quinqueloculina cliarensis
Quinqueloculina lata
Quinqueloculina oblonga
Quinqueloculina seminulum
Reophax scottii
Rosalina anomala

Northern

Buccella frigida
Cribrostomoides jeffreysii
Pateoris hauerinoides

Collection of Samples and their Preservation

The choice of area to be sampled will depend on whether or not living examples are required. If the latter is the case samples should be collected from the lowest part of the intertidal zone at low water or from subtidal areas. In general, muddy sediments will be more productive than clean sands, especially if the latter are moved on each ebb and flow of the tide. Before sampling for living forms it might be worth taking some trial samples to be processed after preservation (see below). Scrape off the top centimetre of sediment over an area of at least 10 × 10 cm. Place it in a jar with water collected from the adjacent channel. Transport it back to the laboratory as quickly as possible preferably in a vacuum flask or keep the sample cool while in transit. It is advisable to measure the water temperature at the time of sampling and to keep the sample one or two degrees below this temperature.

On arrival at the laboratory put the sediment into a clean, shallow dish and allow it to settle out. When this has happened there are two simple methods of trying to isolate the living foraminiferids. Both methods make use of the fact that shallow water foraminiferids are negatively geotropic, i.e. they naturally try to climb upwards.

The first method is to wait for the foraminiferids to climb up the sides of the shallow tray. As they often carry clay minerals and food detritus on their pseudopodial net they may be conspicuous against the tray side. However, they are very small (normally <0·5 mm) and many other, larger and more active organisms also have similar behaviour. With the aid of a spatula and a fine paint brush, it is possible to remove the finds from the tray to a petri dish of filtered water from the collection area and the non-foraminiferal organisms can be separated off when the material is examined under a microscope.

The second method is to pipette off the supernatant fluid once the sediment has settled out and to place a piece of filter paper on the sediment surface. Cover the tray to prevent evaporation and leave for several hours. Then carefully lift off the paper, place it upside down in a petri dish and add a little filtered water from the collection area. In theory, the foraminiferids should have climbed to the sediment surface and adhered to the paper. It may be necessary to make one or two changes of water in the petri dish to clear away any mud which is present. (Full details of the collection of living individuals are given by Arnold, 1974.)

If it is necessary to collect material and preserve it, then the following procedure should be adopted.

(1) Scrape off the top 1 cm of sediment (if a quantitative study is to be made, a known area, such as 10 × 10 cm, should be sampled).

(2) Place the sediment in a bottle and add ethanol in equal volume. Shake vigorously to ensure mixing of preservative and sediment. The sample can then be kept for many months if necessary.

(3) In the laboratory shake the bottle and then pour the contents into a 200 mesh sieve (aperture 76 μm) and wash with water to remove the clay.

(4) When clean, tip the contents of the sieve into a container (bowl) making sure that no sample is lost. Add an equal volume of rose Bengal stain (1 g rose Bengal in 1000 ml distilled water). After 45 min tip the sample into the sieve and

wash again to remove the excess stain. Return the sample to the container and allow to dry (if this is carried out in an oven do not use a temperature >80°C).

(5) When dry allow to cool to room temperature. If the sample was very muddy no further preparation may be necessary. However, if the sample contains significant amounts of silt or sand further concentration of the foraminiferids will be advisable.

(6) *This procedure should only be carried out in a fume chamber and the operator should wear rubber gloves.* The following equipment should be set up in the fume chamber. Clamp a small funnel in a retort stand above a clean empty beaker. Place a filter paper in the funnel. Tip the sample into a separate beaker and pour into it either carbon tetrachloride or trichlorethylene so that there is an excess of liquid. Stir the sample with a glass rod to throw the material into suspension and then allow to settle. The sediment will sink to the bottom while the foraminiferids and other lighter material will float as a scum on the surface of the liquid. Carefully decant this into the filter paper. Collect the used fluid from the beaker beneath the funnel and return it to the sample. Repeat the procedure two or three times until no further scum forms. Then decant all the fluid into the filter paper. Both parts of the sample must be dried in the fume chamber before they are examined. To dry the contents of the beaker it will be necessary to tilt the latter on its side (the vapour is heavier than air). The carbon tetrachloride or trichlorethylene can be reused on other samples.

The principle on which this operation works is that the weight of an air filled foraminiferid test is less than that of the liquid it displaces.

Examination under the Microscope

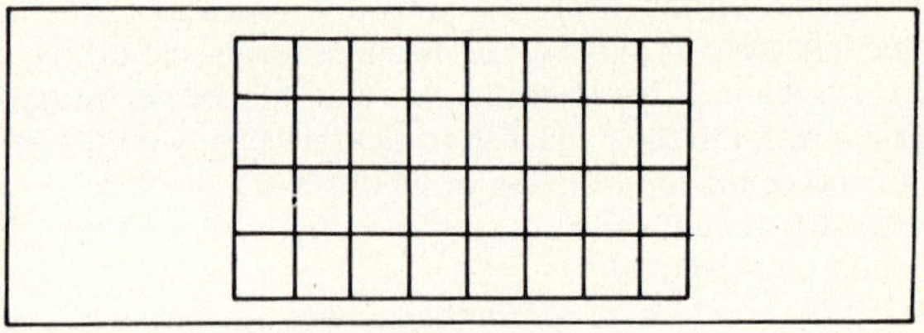

FIG. 2. Cardboard slide. The gridded part is normally black with white lines and it is surrounded by cardboard, thickness 1 mm. Such slides have a glass cover and are held together in a metal sleeve.

Use a biological microscope with transmitted light or dark field illumination to look at living animals. Dead and preserved material is normally examined dry under a stereoscopic binocular microscope, using reflected light. A cardboard slide with a black background makes a useful specimen tray (Fig. 2).

Rose Bengal stains protoplasm bright red so forms which contained protoplasm at the time of collection (and therefore presumed to have been living) are easily differentiated from unstained empty tests.

Specimens are normally mounted on slides of the type illustrated in Fig. 2. Paint the surface of the slide with dilute water-soluble gum and allow to dry. When a moistened foraminiferid is placed on the slide it will automatically stick. Specimens may be removed or changed in orientation using a moistened paintbrush (size 000 or 00 sable).

Preparation of the Material to Examine the Wall Structure

As the basic classification into suborders is dependent on test wall composition and structure it is essential that this should be determined properly.

The simplest method of looking at wall structure is to mount specimens on a glass slide, surround them with some broken pieces of cover glass, place a drop of runny canada balsam on a cover glass, and then gently lower it onto the specimens. Allow it to stand at room temperature for several days and normally by this time the balsam will have penetrated into most if not all the chambers of the foraminiferid.

If it is desired to make a thin section of a foraminiferid the following procedure should be carried out.

Place a small piece of "Lakeside" cement on a glass slide and warm it on a hotplate. Allow to cool and place the foraminiferid on the "Lakeside".* Warm again to allow the foraminiferid to settle into the cement. If a particular orientation of the foraminiferid is required it can be manipulated with a heated needle.

Grinding is carried out under a microscope. Place a drop of water on the "Lakeside" and, using a ready-prepared ground glass slide, rub away at the foraminiferid. When a flat surface has been prepared use a heated needle to turn the specimen over so that the flat side now faces downwards. Grinding is recommenced and continued until the section is of the desired thickness. The section may then be covered with balsam and a cover glass.

*"Lakeside" cement can be obtained from Whithear Lapidary, 35 Manor Way, North Harrow, U.K.

Recognition of Wall Composition and Structure

Modern foraminiferids fall into two main groups based on differences of composition: agglutinated and calcareous.

Agglutinated walls, otherwise known as arenaceous, are made up of detrital grains glued together with cement (Fig. 3A, B). The grains are commonly quartz and often of silt size but they may be coarser. There may also be various other minerals and even biogenic material such as sponge spicules and shell debris. The cement is basically organic (tectin and acid mucopolysaccharide), sometimes with an iron content which imparts a rusty or reddish colour. Calcareous cement may also be present although this is never developed in brackish water foraminiferids.

Agglutinated walls are easily recognized from their content of detrital particles which are normally obvious in solid specimens, in a balsam mount and in thin section. Such walls characterize the suborder Textulariina.

Calcareous walls are composed of calcium carbonate predominantly as the mineral calcite. Two main types of structure are known, porcellaneous and hyaline.

The porcellaneous wall structure is made up of tiny crystallites of calcite in a mainly randomly arranged fashion (Fig. 3C). The thickness of the wall includes numerous crystallites. No pores are present and the wall normally appears translucent or opaque and white or buff. In balsam mounts or sections the wall is fine grained and amorphous in appearance. Walls of this kind characterize the suborder Miliolina.

Hyaline means glassy and in a fresh state this describes the appearance of such walls. Here the calcite crystallites are arranged either in a radial pattern (Fig. 3E, F) or with the *c*-axis oblique to the wall surface (granular) (Fig. 3D), or rarely the whole test behaves as a single crystal. Hyaline walls always contain pores although sometimes these are too small to see at low magnification. In balsam mounts hyaline walls appear transparent. Under crossed polarisors the radial varieties show an extinction cross in each chamber. In granular varieties the crystals extinguish randomly as the specimen is rotated under crossed polarisors. The single crystal forms show total extinction. The same pattern can be seen in thin section. In addition it will be noticed that the wall has a layered structure in all but the final chamber. Walls of this kind characterize the suborder Rotaliina. Rarely, representatives of the suborder Rotaliina have walls made of aragonite. This may be distinguished from calcite by means of a staining test in which aragonite turns black and calcite remains unstained. The stain is made up as follows: dissolve 11·9 g of manganese sulphate in 100 ml distilled water, then add 1 g of silver sulphate. Allow it to cool, filter it and add two drops of 10% sodium hydroxide solution. Allow to stand for two hours, then filter. (The solution should be stored in a dark bottle.) Staining is carried out at room temperature and a reaction should be produced in a matter of minutes.

In muddy sediments, low pH conditions are often developed beneath the surface veneer of oxidized material. Under these conditions calcareous tests undergo dissolution. In the early stages their surface becomes etched and this causes porcellaneous and hyaline walls to become opaque. Caution must therefore be exercised in the recognition of wall structures.

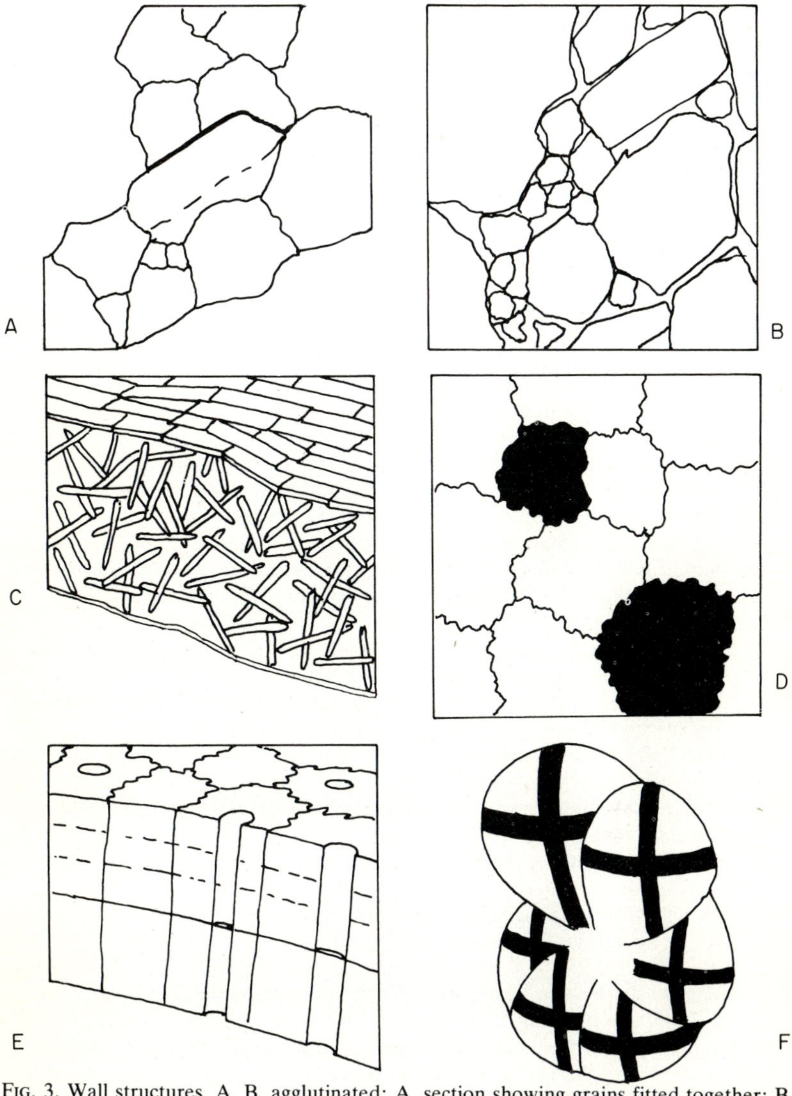

FIG. 3. Wall structures. A, B, agglutinated; A, section showing grains fitted together; B, surface view. C, porcellaneous, outer surface layer of laths over a jumble of calcite needles. D, surface view of a calcareous hyaline granular wall under crossed polarisors showing two crystals in extinction and the remainder not in extinction. E, block view of a calcareous hyaline radial wall showing crystal units, pores and layers within the wall. F, a calcareous hyaline radial foraminiferid under crossed polarisors showing extinction crosses, one on each chamber.

General Description of the Test including a Glossary of Terms Used

The foraminiferid shell is normally referred to as the *test*, because, in some forms at least, the test is within the protoplasm rather than outside it. In its simplest form the test consists of a single chamber in which there is a single opening, the aperture. However, the majority of genera are more complex, being made up of several chambers arranged according to the pattern of growth. Figure 4 shows an example of a foraminiferid with the basic morphological features labelled. These are described in greater detail below.

PROLOCULUS: The first-formed chamber.

UNISERIAL: The chambers arranged in a single row (Fig. 5B).

BISERIAL: Two columns of chambers added on alternate sides of the test (Fig. 12D).

TRISERIAL: Three columns of chambers added in succession; a high trochospiral with three chambers per whorl (Fig. 6A).

PLANISPIRAL: Chambers added in a single plane about an axis of coiling. *Evolute* if all the chambers are visible on both sides (Fig. 7B). *Involute* if the last whorl covers the earlier whorls (Fig. 5G).

TROCHOSPIRAL: Chambers added in a spiral with translation along the axis of coiling (as in a snail shell). The spiral side is evolute and the umbical side involute (Fig. 6E–G). Coiling to left or to right is of no specific significance in benthic foraminiferids.

SPIROLOCULINE: Each chamber occupies one half whorl, planispiral, all chambers visible from each side (Fig. 7D–G).

TRILOCULINE: Each chamber is elongate and occupies one half whorl, successive chambers being added in planes 120° apart (Fig. 10F–H).

QUINQUELOCULINE: As in triloculine but the planes being 144° apart (Fig. 8D–F).

MILIOLINE: A general term to include triloculine and quinqueloculine and intermediate forms (Fig. 10F–H).

CYCLICAL: Chambers added in more or less circular or concentric patterns. (Fig. 21D, E).

SEPTUM: (plural septa) Internal wall separating adjacent chambers.

SUTURE: The surface expression of a septum.

RETRAL PROCESS: An arch-like bridge running from the back of one chamber to the front of the previous chamber, spanning the suture.

APERTURE: An opening from the chamber to the outside. It may be just a circular opening without any additional features, in which case it is said to be *simple*. If it is at the end of the test it is *terminal*. It may be surrounded by a thickening—*lip*, partly covered by a *flap*, or be partly filled by an outgrowth—*tooth*. Sometimes an internal *toothplate* is present—a wall which runs within the chamber to the aperture. A few species have a tube leading from the aperture into the cavity of the chamber—*entosolenian tube*.

UMBILICUS: (plural umbilici) Depression at the centre of a spiral, around the coiling axis.

ORNAMENT: Features developed on the test surface; *tubercles*—small granular outgrowths; *bosses*—massive outgrowths; *ribs*—enlongate outgrowths; *keels*—peripheral thickenings.

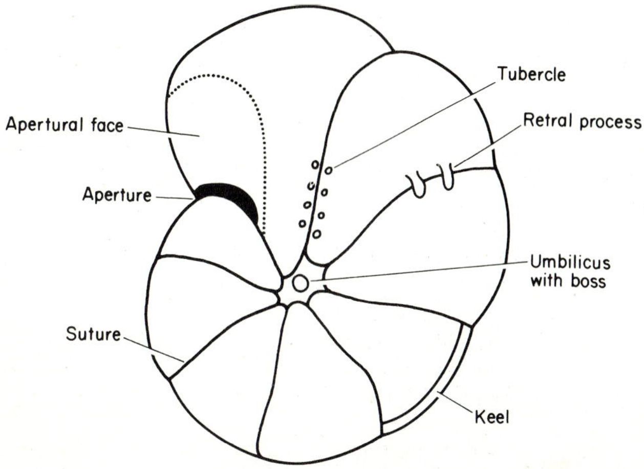

FIG. 4. Diagram to show some general morphological features of the test.

Classification

The classification is that of Loeblich and Tappan (1964) with modifications proposed in 1974.

Phylum PROTOZOA
- Subphylum SARCODINA
  - Class RHIZOPODEA
    - Subclass GRANULORETICULOSIA
      - Order FORAMINIFERIDA
        - Suborder TEXTULARIINA
          - Superfamily Lituolacea
            - Family Hormosinidae
              - *Reophax moniliformis*
              - *Reophax scottii*
            - Family Rzehakinidae
              - *Miliammina fusca*
            - Family Lituolidae
              - *Haplophragmoides wilberti*
              - *Cribrostomoides jeffreysii*
              - *Ammobaculites balkwilli*
            - Family Trochamminidae
              - *Trochammina inflata*
              - *Trochammina ochracea*
              - *Jadammina macrescens*
            - Family Ataxophragmiidae
              - *Eggerelloides scaber*
              - *Clavulina obscura*
        - Suborder MILIOLINA
          - Superfamily Miliolacea
            - Family Fischerinidae
              - *Cyclogyra involvens*
            - Family Nubeculariidae
              - *Cornuloculina balkwilli*
              - *Spiroloculina excavata*
              - *Spiroculina rotunda*
            - Family Miliolidae
              - *Quinqueloculina bicornis*
              - *Quinqueloculina bicornis angulata*
              - *Quinqueloculina cliarensis*
              - *Quinqueloculina dimidiata*
              - *Quinqueloculina lata*
              - *Quinqueloculina oblonga*
              - *Quinqueloculina seminulum*
              - *Miliolinella subrotunda*

Suborder ROTALIINA
  Superfamily Nodosariacea
    Family Nodosariidae
      Lagena spp.
      Oolina spp.
      Fissurina lucida
      Fissurina marginata
  Superfamily Buliminacea
    Family Turrilinidae
      Buliminella elegantissima
    Family Bolivinitidae
      Bolivina pseudoplicata
      Brizalina pseudopunctata
      Brizalina variabilis
    Family Buliminidae
      Bulimina marginata
      Trifarina angulosa
  Superfamily Discorbacea
    Family Discorbidae
      Buccella frigida
      Gavelinopsis praegeri
      Rosalina anomala
      Rosalina williamsoni
    Family Glabratellidae
      Galbratella millettii
    Family Asterigerinidae
      Asterigerinata mamilla
    Family Cibicididae
      Cibicides lobatulus
    Family Planorbulinidae
      Planorbulina mediterranensis
    Family Acervulinidae
      Acervulina inhaerens
  Superfamily Spirillinacea
    Family Spirillinidae
      Spirillina vivipara
      Patellina corrugata
  Superfamily Rotaliacea
    Family Rotaliidae
      Ammonia beccarii
    Family Elphidiidae
      Elphidium crispum
      Elphidium cuvillieri
      Elphidium earlandi
      Elphidium excavatum
      Elphidium gerthi

Elphidium incertum
Elphidium macellum
Elphidium margaritaceum
Elphidium oceanensis
Elphidium williamsoni
Protelphidium germanicum
Superfamily Cassidulinacea
Family Caucasinidae
Fursenkoina fusiformis
Family Cassidulinidae
Cassidulina obtusa
Superfamily Nonionacea
Family Nonionidae
Nonion depressulus
Superfamily Robertinacea
Family Ceratobuliminidae
Lamarckina haliotidea

Key to Genera

This key is designed to guide the reader to the genera of British brackish-water and nearshore foraminiferids. Where more than one species of one genus are known the distinctive differences are discussed under the genus.

The key starts with a basic division, based on wall composition, into suborders. Beyond that the key does not adhere to the currently accepted classification of Loeblich and Tappan. Instead, it lists the simplest diagnostic features by which genera may be recognized. Because of this simplicity the key should not be extended in its usage to open marine foraminiferids which are, of course, much more diverse than those of estuaries and nearshore areas.

As far as possible the descriptive terms have been kept simple. Definitions of other terms are given on p. 14 and a proper understanding of them is essential to the proper usage of the key.

Identification is normally made on dry specimens. Material mounted in any liquid will be difficult to identify.

1. Wall composition:

Calcareous **2**

Agglutinated .... Suborder TEXTULARIINA **3**

2. Wall without pores, porcellaneous .... Suborder MILIOLINA **6**

Wall with pores, hyaline .... Suborder ROTALIINA **9**

3. Suborder TEXTULARIINA. Growth form:

Single row of chambers (uniserial) .... *Reophax* (p. 23)

Triserial then uniserial .... *Clavulina* (p. 26)

Coiling milioline .... *Miliammina* (p. 23)

Planispirally coiled throughout **4**

Planispirally coiled then uncoiled .... *Ammobaculities* (p. 24)

Trochospirally coiled then triserial .... *Eggerelloides* (p. 26)

Trochospirally coiled throughout **5**

4. Aperture at base of last chamber .... *Haplophragmoides* (p. 24)

Aperture on face of last chamber .... *Cribrostomoides* (p. 24)

5. Aperture on final chamber

Single slit .... *Trochammina* (p. 26, 28)

Slit plus additional circular openings .... *Jadammina* (p. 28)

6. Suborder MILIOLINA. Growth form:

Proloculus then long planispirally coiled chamber.... *Cyclogyra* (p. 29)

Several chambers coiled planispirally .... *Cornuloculina* (p. 29)

Spiroloculine coiling throughout .... *Spiroloculina* (p. 29)

Quinqueloculine coiling throughout **7**

Initially quinqueloculine then spiroloculine.. **8**

Triloculine coiling with apertural flap .... *Miliolinella* (p. 36)

7. Aperture

With tooth .... *Quinqueloculina* (p. 30)

With flap .... *Miliolinella* (p. 36)

8. Chambers compressed, aperture with tooth .... *Massilina* (p. 36)

Chambers inflated, aperture without tooth or flap .... *Pateoris* (p. 36)

9. Suborder ROTALIINA

Wall of single crystal of calcite **10**

Wall of many crystallites of calcite **11**

Wall of aragonite .... *Lamarckina* (p. 64)

10. Enrolled non-septate (i.e. undivided) tube .... *Spirillina* (p. 38)

Trochospirally coiled tube with incomplete secondary septa
Patellina (p. 38)

11. Single chamber **12**

Several chambers **14**

12. Aperture

On a neck with lip .... *Lagena* (p. 38)

With entosolenian tube **13**

13. Chamber

Globular .... *Oolina* (p. 40)

Compressed .... *Fissurina* (p. 40)

14. Chamber arrangement

Biserial in a single plane **15**

Twisted biserial .... *Fursenkoina* (p. 41)

Enrolled biserial .... *Cassidulina* (p. 64)

Triserial **16**

Planispiral **17**

High trochospiral .... *Buliminella* (p. 41)

Low trochospiral throughout **19**

Low trochospiral in early stages, becoming annular or irregular **23**

15. Sutures

Straight or curved .... *Brizalina* (p. 41, 42)

Crenulated with retral processes .... *Bolivina* (p. 42)

16. Cross section rounded, aperture comma shaped .... *Bulimina* (p. 42)

Cross section triangular, aperture circular .... *Trifarina* (p. 42)

17. Sutures bridged by retral processes .... *Elphidium* (p. 44)

Sutures without retral processes **18**

18. Wall structure radial .... *Protelphidium* (p. 54)

Wall structure granular .... *Nonion* (p. 54)

19. Test biconvex **20**

Test plano-convex **21**

Test concavo-convex .... *Rosalina* (p. 58)

20. Umbilicus obscured by granulations .... *Buccella* (p. 54)

Umbilicus without granules, with or without umbilical boss
Ammonia (p. 56)

21. Umbilical side flat **22**

Spiral side flat .... *Cibicides* (p. 62)

22. Umbilicus with large flat boss .... *Gavelinopsis* (p. 58)

Umbilicus without boss, radial ornament on umbilical side
Glabratella (p. 60)

Umbilicus closed and surrounded by secondary chambers
Asterigerinata (p. 60)

23. Adult growth cyclical, compressed.. *Planorbulina* (p. 62)

Adult growth irregular, chambers inflated .... *Acervulina* (p. 62)

Systematic Descriptions

Suborder TEXTULARIINA

Reophax moniliformis Siddall

(Fig. 5A, B)

Wall agglutinated, composed of detrital grains and/or sponge spicules with an organic cement. Test cylindrical, consisting of a single column of chambers. Sutures depressed. Aperture a simple terminal hole. Average length 0·8 mm. A sediment dwelling, inner shelf, essentially marine species which sometimes penetrates into the mouths of estuaries.

Reophax scottii Chaster

(Fig. 5C)

Wall agglutinated, composed of detrital mica flakes arranged in an overlapping series like tiles on a roof, held together by an organic cement. Test uniserial, flexible at the deeply incised sutures. Aperture a simple terminal hole. Average length 0·7 mm. A muddy sediment, inner shelf marine species.

Miliammina fusca (Brady)

(Fig. 5D–F)

Wall agglutinated, composed of silt-sized detrital grains in an organic cement, pale brown in colour. Test elongate, rounded in section, made up of many chambers coiled on a milioline plan. Sutures slightly depressed. Aperture, with a small tooth. Average length 0·4 mm. A sediment dwelling euryhaline species characteristic of brackish lagoons, estuaries and tidal marshes. It has a recorded salinity tolerance range of 0–35 ‰.

Haplophragmoides wilberti Anderson

(Fig. 5G, H)

Test agglutinated, composed of very fine silt grains in an organic cement, pale brown in colour and smoothly finished. Test planispirally coiled with seven to nine chambers visible in the last whorl. Sutures distinct, slightly depressed, straight or sigmoid. Aperture a slit at the base of the final chamber. Average maximum diameter 0·4 mm. A sediment dwelling, high marsh species recorded only from the Dovey estuary.

Cribrostomoides jeffreysii (Williamson)

(Fig. 5I, J)

Wall agglutinated, composed of fine silt grains in an organic cement, pale brown in colour and smoothly finished. Test planispirally coiled with six to seven chambers in the last whorl. Sutures distinct, depressed, slightly sigmoid. Aperture a lipped slit just above the base of the apertural face. Average maximum diameter 0·5 mm. A marine inner shelf species.

Ammobaculites balkwilli Haynes

(Fig. 5K)

Test agglutinated, composed of coarse angular quartz grains in a white matrix of fine silt, with scattered grains of other minerals including coal debris. Test starts planispirally coiled but then becomes uncoiled. Normally three to five chambers visible in the planispiral portion and three to five chambers in the uncoiled part. The latter chambers vary greatly in shape but generally the uncoiled part is straight and cylindrical. Sutures depressed. Aperture large, round and terminal. Average length 0·7 mm. A sediment dwelling species abundant in the marshes of the Dovey estuary, and not so far recorded from elsewhere.

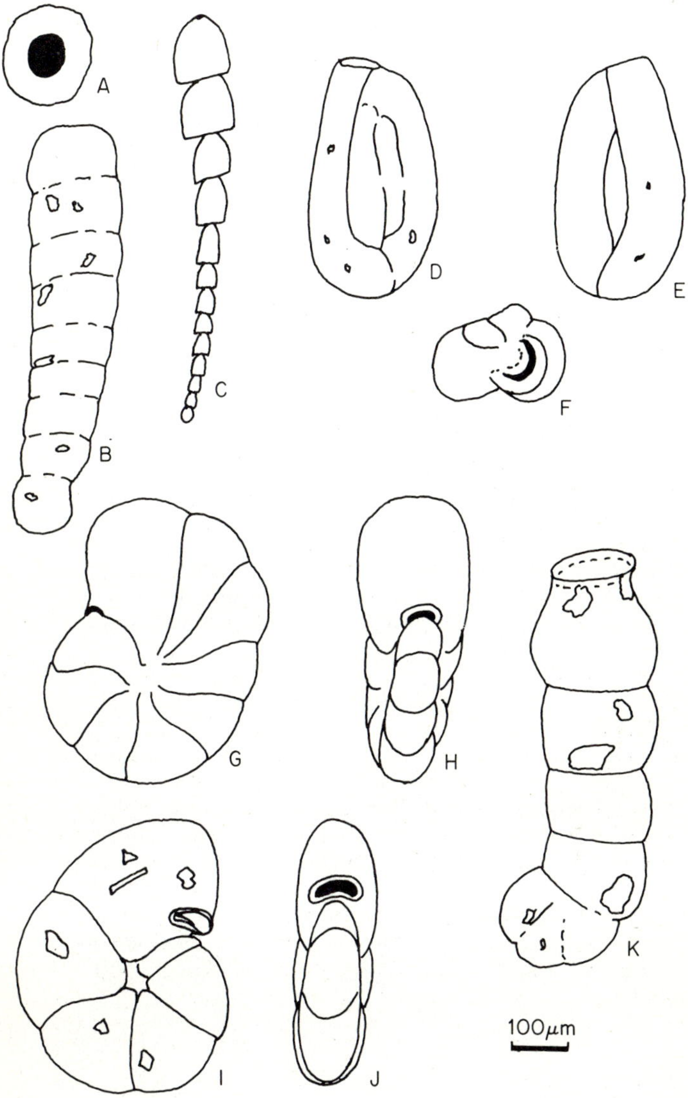

FIG. 5. Suborder Textulariina—wall agglutinated. A, B, *Reophax moniliformis* Siddall. C, *Reophax scotti* Chaster. D, E, F, *Miliammina fusca* Brady. G, H, *Halophragmoides wilberti* Anderson. I, J. *Cribrostomoides jeffreysii* (Williamson). K, *Ammobaculites balkwilli* Haynes.

Eggerelloides scaber Haynes

(Fig. 6A–C)

Eggerella scabra (Williamson) of authors

Test agglutinated, consisting of detrital grains of variable size in an organic cement, white to pale brown in colour. Initial part of test trochospirally coiled but adult part is triserial. Chambers increase rapidly in size so that the last whorl commonly forms half the length of the test. Sutures depressed. Aperture a high arch at the basal suture of the last chamber. Average length 0·6 mm. A sediment dwelling essentially marine, inner shelf species (tolerates a salinity range of 29–35‰), widely distributed along the coasts of northern Europe, and penetrating into the mouths of estuaries.

Clavulina obscura Chaster

(Fig. 6D)

Wall agglutinated, composed of detrital particles in an organic cement, roughly finished. Early part of test triserial but later uniserial. Chambers inflated with deeply depressed sutures. Aperture a terminal hole bordered by a rim of cement. Average length 0·4 mm. An inner shelf, marine species.

Trochammina inflata (Montagu)

(Fig. 6E–G)

Test agglutinated, composed of extremely fine detrital grains with abundant organic cement, brown in colour. Test trochospirally coiled with globular chambers. On the spiral side the sutures are gently depressed. On the umbilical side the sutures are deeply depressed between the inflated chambers. Umbilicus deep. Aperture a narrow slit with bordering lip. Average greatest diameter 0·4 mm. A sediment dwelling species typical of high salt marsh, able to withstand great variation in the environment including periods of low oxygen availability.

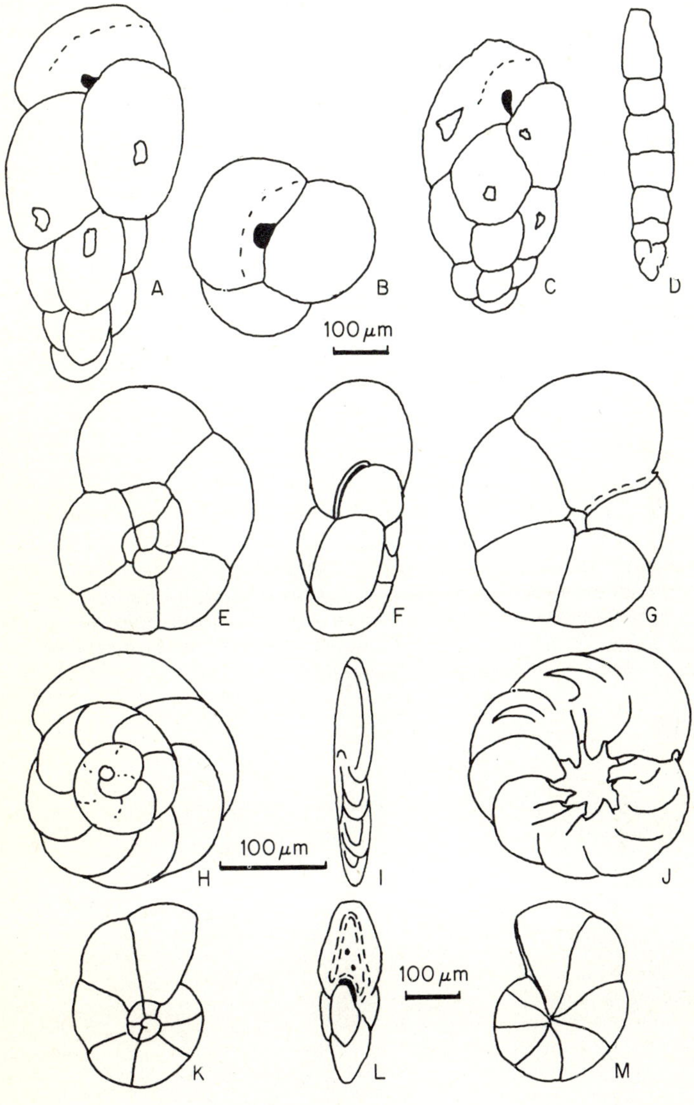

Fig. 6. Suborder Textulariina—wall agglutinated. A, B, C, *Eggerelloides scaber* Haynes. D, *Clavulina obscura* Chaster. E, F, G, *Trochammina inflata* (Montagu). H, I, J. *Trochammina ochracea* (Williamson). K, L, M, *Jadammina macrescens* (Brady). The apertural view, L, is of a partially collapsed specimen.

Trochammina ochracea (Williamson)

(Fig. 6H–J)

Wall agglutinated, composed of very fine detritus in an organic cement, reddish-brown in colour. Test trochospirally coiled, very compressed, with acute periphery. On the spiral side the sutures are very slightly depressed and strongly curved backward. On the slightly concave umbilical side the sutures are depressed and arcuate. Umbilicus wide, shallow and star-shaped between umbilical chamber extensions. Aperture an arch at periphery with secondary openings into the umbilicus. Average maximum diameter 0·2 mm. An inner shelf species which lives clinging to pebbles, seaweeds, etc.

Jadammina macrescens (Brady)

(Fig. 6K–M)

Test agglutinated, composed of extremely fine detrital grains with abundant organic cement, brown in colour. The wall is thin and flexible when wet and the chambers commonly collapse when dried. Test trochospirally coiled, compressed with shallow umbilicus. Sutures slightly depressed (but when the chambers collapse they appear raised). Aperture a slit at the base of the apertural face and with accessory circular openings scattered over the face. Average greatest diameter 0·3 mm. Common on brackish marshes throughout the world, and able to withstand extreme environmental variability.

Suborder MILIOLINA

Cyclogyra involvens (Reuss)

(Fig. 7A, B)

Wall calcitic, porcellaneous, white. Test starts with a proloculus followed by a long, planispirally coiled, tubular chamber. Aperture terminal. Average maximum diameter 0·2 mm. An inner shelf marine species, often found on pebbles and seaweeds.

Cornuloculina balkwilli (Macfadyen)

(Fig. 7C)

Wall calcitic, porcellaneous, white. Test oval in outline, compressed, each chamber forms about two thirds of a whorl, aperture simple, terminal. Average length 0·15 mm. Probably an inner shelf, marine species transported into the mouths of muddy estuaries (e.g. Severn).

Spiroloculina excavata d'Orbigny

(Fig. 7D, E)

Wall calcitic, porcellaneous, white. Test coiled on a spiroloculine plan, each chamber spanning a half whorl. Laterally compressed with rectangular periphery. Aperture terminal, with a small tooth. Average length 0·7 mm. An inner shelf, marine species.

Spiroloculina rotunda d'Orbigny

(Fig. 7F, G)

Wall calcitic, porcellaneous, white to brown, especially along the sutures. Test coiled on a spiroloculine plan, each chamber spanning a half whorl. Umbilical regions depressed, periphery rounded. Aperture terminal, with a tooth. Average length 0·6 mm. An inner shelf, marine species.

Quinqueloculina

Wall porcellaneous, translucent to opaque. Chambers coiled on a quinqueloculine plan so that five are visible externally, four on one side (Fig. 8D) and three on the other (Fig. 8F), two being common to both views. Aperture terminal, usually with a tooth.

Key to species of *Quinqueloculina*

1. Test ornamented with ribs **2**
Test smooth **3**

2. Section rounded .... *bicornis* (p. 30)
Section angular .... *bicornis* var. *angulata* (p. 32)

3. Section triangular .... *cliarensis* (p. 32)
Section rounded **4**

4. Outline elongate .... *oblonga* (p. 34)
Outline oblong .... *lata* (p. 34)
Outline oval **5**

5. Average length 0·3 mm .... *dimidiata* (p. 32)
Average length 0·7 mm .... *seminulum* (p. 34)

Quinqueloculina bicornis (Walker and Jacob)

(Fig. 7H–J)

Wall calcitic, porcellaneous, white, ornamented with longitudinal ribs. Test coiled on a quinqueloculine plan. Aperture an elongate slit with a strong tooth. Average length 0·7 mm. An inner shelf, marine species.

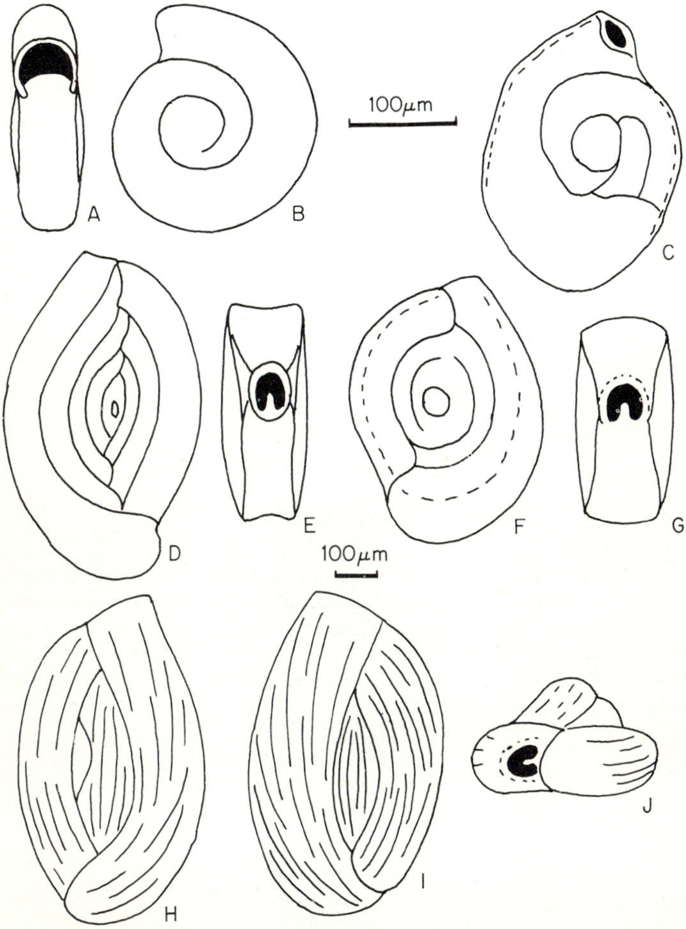

FIG. 7. Suborder Miliolina—wall porcellaneous, white and imperforate. A, B, *Cyclogyra involvens* Reuss. C, *Cornuloculina balkwilli* (Macfadyen). D, E, *Spiroloculina excavata* d'Orbigny. F, G, *Spiroloculina rotunda* d'Orbigny. H, I, J, *Quinqueloculina bicornis* (Walker and Jacob).

Quinqueloculina bicornis (Walker and Jacob) var. *angulata* (Williamson)

(Fig. 8A–C)

Differs from *Q. bicornis sensu stricto* in having angular margins to the chambers and in having less uniformly developed ribbing. Average length 1·5 mm. An inner shelf, marine species.

Quinqueloculina cliarensis (Heron-Allen and Earland)

(Fig. 8D–F)

Wall calcitic, porcellaneous, white. Test coiled on a quinqueloculine plan. Periphery angular, chambers arcuate, last one having a neck. Aperture terminal, with a small tooth. Average length 0·6 mm. An inner shelf, marine species.

Quinqueloculina dimidiata Terquem

(Fig. 8G–I)

Wall calcitic, porcellaneous, translucent to white. Test oval in outline and subcircular in section. Coiling quinqueloculine. Sutures oblique. Aperture a terminal arch generally without a tooth. Average length 0·3 mm. A sediment dwelling, essentially marine species which colonizes the mouths of estuaries when conditions are favourable (usually during the late summer–early autumn).

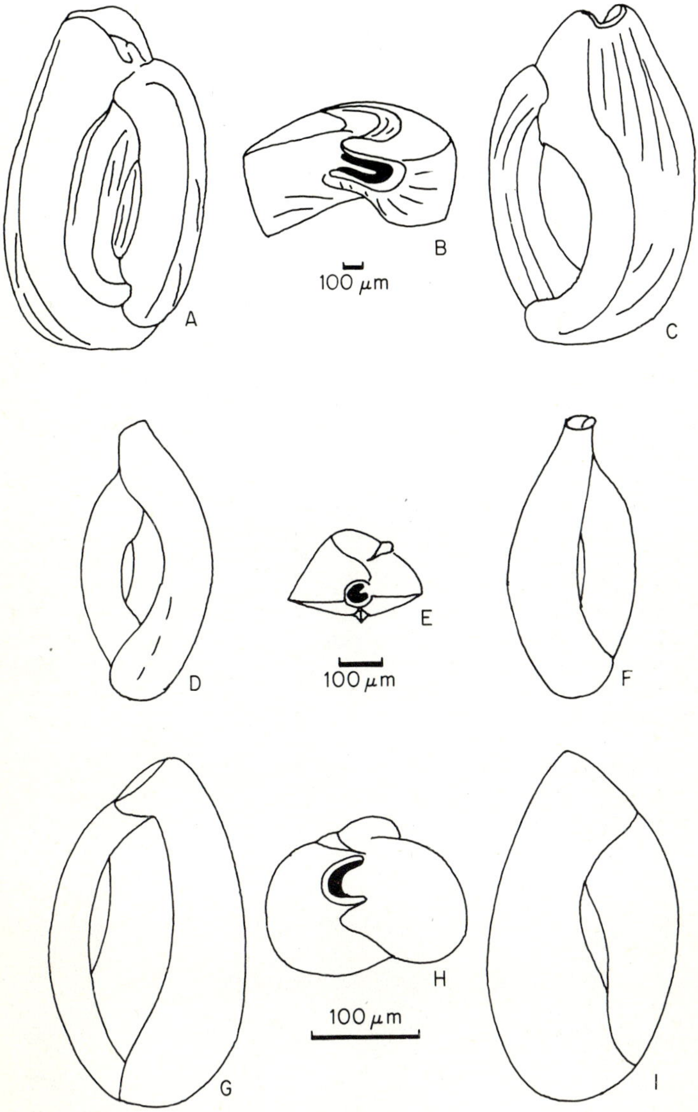

FIG. 8. Suborder Miliolina—wall porcellaneous, white and imperforate. A, B, C, *Quinqueloculina bicornis* (Walker and Jacob) var. *angulata* (Williamson). D, E, F, *Quinqueloculina cliarensis* (Heron-Allen and Earland). G, H, I, *Quinqueloculina dimidiata* Terquem.

Quinqueloculina lata Terquem

(Fig. 9A–C)

Wall calcitic, porcellaneous, white. Test coiled on a quinqueloculine plan. Test outline oblong, cross section rounded, triangular, sutures only slightly depressed. Aperture terminal, rounded, with a small tooth. Average length 0·7 mm. An inner shelf, marine species.

Quinqueloculina oblonga (Montagu)

(Fig. 9D–F)

Wall calcitic, porcellaneous, white. Test coiled on a quinqueloculine to sometimes triloculine plan. Test very elongate, oval in section. Aperture terminal with a tooth (which is rarely flap-like). Average length 0·4 mm. An inner shelf, marine species.

Quinqueloculina seminulum (Linné)

(Fig. 9G–I)

Wall calcitic, porcellaneous, white to buff. Test coiled on a quinqueloculine plan. Test elongate to subrounded in outline, periphery round. Aperture terminal with a tooth. Average length 0·8 mm. An inner shelf, marine species. Very common.

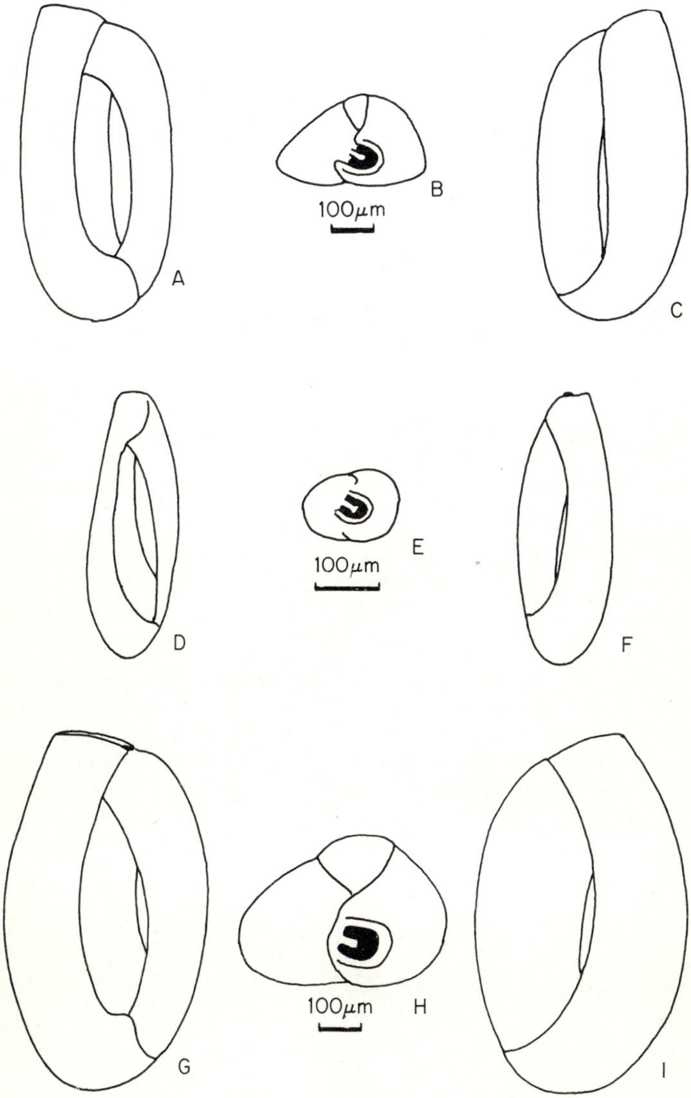

FIG. 9. Suborder Miliolina—wall porcellaneous, white and imperforate. A, B, C, *Quinqueloculina lata* Terquem. D, E, F, *Quinqueloculina oblonga* (Montagu). G, H, I, *Quinqueloculina seminulum* (Linné).

Massilina secans (d'Orbigny)

(Fig. 10A–C)

Wall calcareous, porcellaneous. Test initially coiled on a quinqueloculine plan but later becomes spiroloculine with two chambers per whorl. Test compressed with subangular periphery. Aperture terminal, with large bifid tooth. Average length 1·8 mm. An inner shelf, marine, species commonly found in association with seaweeds.

Pateoris hauerinoides (Rhumbler)

(Fig. 10D–E)

Wall calcite, porcellaneous, brown to white. Test initially coiled on a quinqueloculine plan but later becomes spiroloculine with two chambers per whorl. Chambers inflated, periphery rounded. Aperture terminal without a flap or tooth. Average greatest length 0·4 mm. An inner shelf, marine species.

Miliolinella subrotunda (Montagu)

(Fig. 10F–H)

Wall calcitic, porcellaneous, white. Test circular in outline and subcircular in section. Coiling mainly quinqueloculine although triloculine variants are known. Aperture a terminal arch with a small flap. Average length 0·2 mm. Lives on sediment and on weeds, essentially marine but colonizes the mouths of estuaries when conditions are favourable.

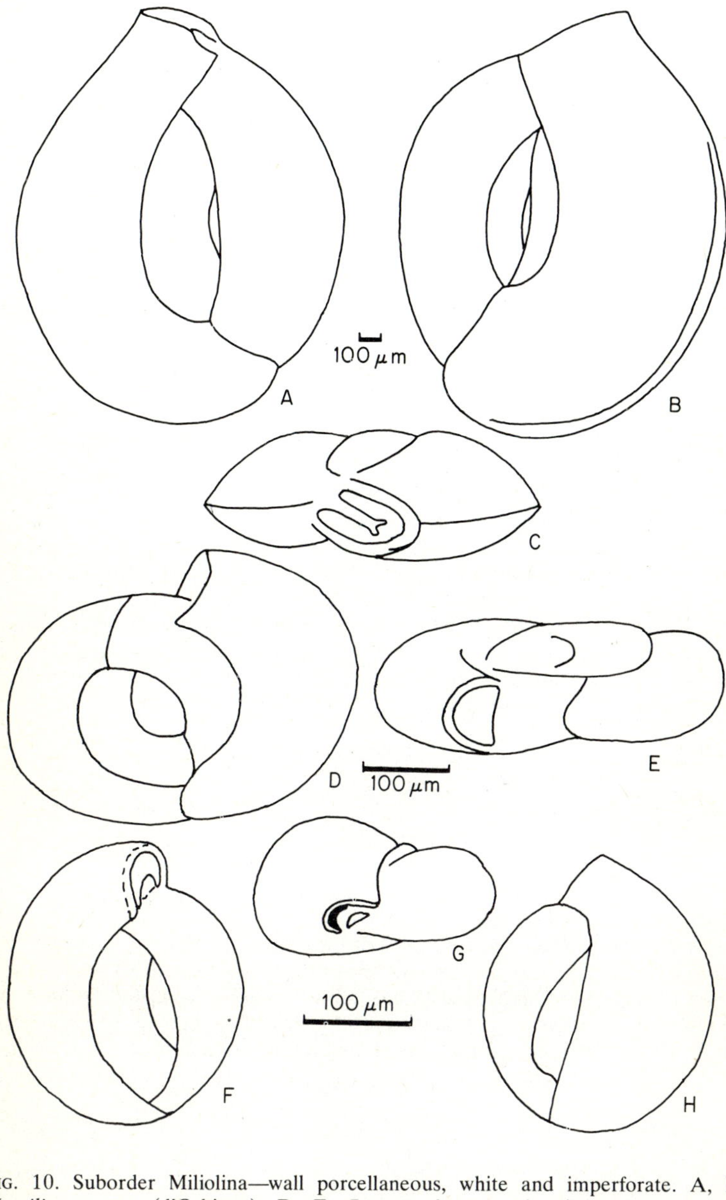

FIG. 10. Suborder Miliolina—wall porcellaneous, white and imperforate. A, B, C, *Massilina secans* (d'Orbigny). D, E, *Pateoris hauerinoides* (Rhumbler). F, G, H, *Miliolinella subrotunda* (Montagu).

Suborder ROTALIINA

Spirillina vivipara Ehrenberg

(Fig. 11A)

Wall calcitic, behaves optically as a single crystal, glassy and with pores. Test planispirally coiled, compressed, consisting of the proloculus and a long tubular chamber which is undivided. Aperture terminal, simple. Average greatest diameter 0·1 mm. A marine, inner shelf species which lives clinging to firm substrates such as pebbles and seaweeds. Sometimes found in the mouths of estuaries.

Patellina corrugata Williamson

(Fig. 11B–D)

Wall calcitic, behaves optically as a single crystal, glassy and with pores. Test is cone-shaped (planoconvex), trochospirally coiled. The proloculus is followed by a tubular chamber of one to three whorls and then by chambers each making one half whorl. On the spiral side the sutures are thickened and raised. Around the periphery on the umbilical side is a row of deep pits. The aperture opens beneath a flap. Average greatest diameter 0·3 mm. A marine, inner shelf species which lives clinging to firm substrates such as pebbles and seaweeds. Sometimes found in the mouths of estuaries.

Genus *Lagena*

(Fig. 11E)

Wall calcitic, of radially arranged crystallites with very fine pores, often transparent. Test single-chambered, of variable form, but always circular in cross section, and with an elongate neck at the end of which is a round aperture bordered by a lip. The test may be smooth or ornamented with striae, ribs, etc. Average length 0·3–0·6 mm. Species are numerous and recognized on the form of the test and the nature of the ornament. A marine, shelf genus which lives on muddy substrates. Postmortem transport during storms sometimes leads to their deposition in estuarine muds.

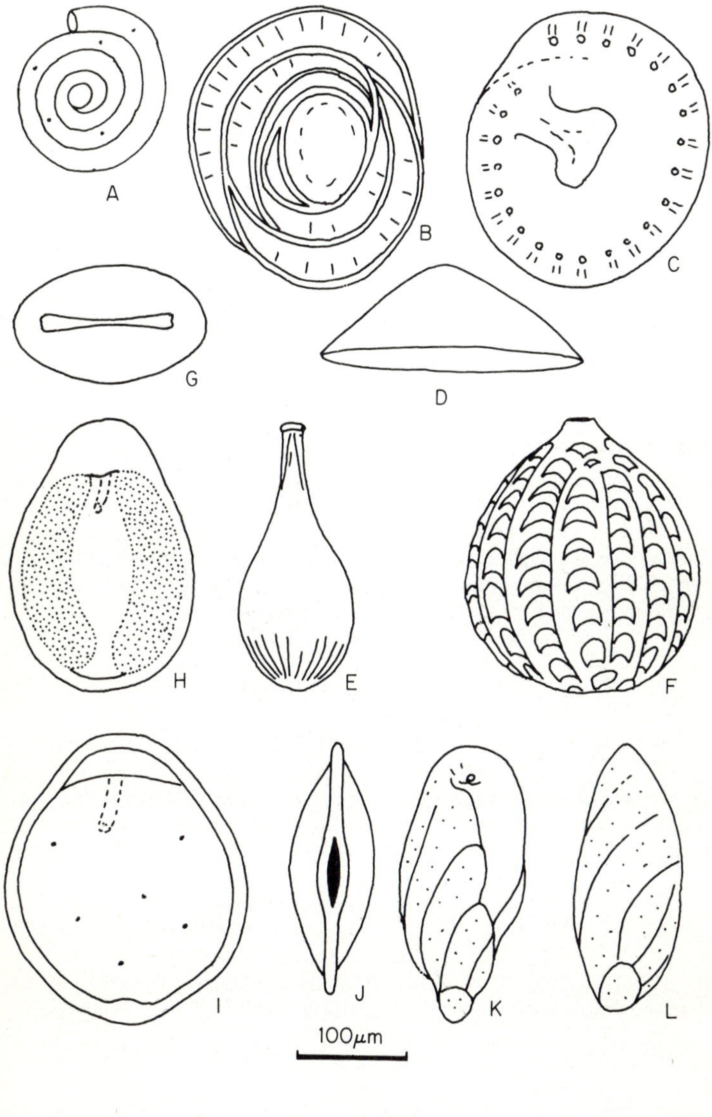

FIG. 11. Suborder Rotaliina—wall hyaline, perforate. A, *Spirillina vivipara* Ehrenberg. B, C, D, *Patellina corrugata* Williamson. E, *Lagena semistriata* Williamson. F, *Oolina melo* d'Orbigny. G, H, *Fissurina lucida* (Williamson). I, J, *Fissurina marginata* (Montagu). K, L, *Buliminella elegantissima* (d'Orbigny).

Genus *Oolina*

(Fig. 11F)

Wall calcitic, of radially arranged crystallites, with very fine pores. Test single-chambered, spherical to ovoid, always circular in cross section. Aperture a round opening, an entosolenian tube extends from the aperture into the chamber of the test. Ornament takes the form of ribs, or other narrow raised features which may be organized into hexagonal or other patterns. Average length 0·2–0·3 mm. Several species are recognized on variation in test form and the nature of the ornament. A marine, shelf genus, normally commonest on muddy substrates. Some species may be parasitic on other foraminiferids. Reproduction may be exclusively asexual and the single small nucleus may be haploid.

Genus *Fissurina*

Wall calcitic, of radially arranged crystallites, with very fine pores, often transparent. Test single-chambered, oval in outline, compressed, never circular in cross section. Aperture a lenticular slit in the periphery, with an entosolenian tube extending into the chamber of the test. Ornament consists of peripheral thickenings or translucent to opaque bands. Average length 0·1–0·3 mm. A marine shelf genus, commonest on muddy substrates. May be transported into estuaries and deposited as empty tests in mud.

Fissurina lucida (Williamson)

(Fig. 11G, H)

Characterized by its oval cross section, absence of peripheral thickening and development of a horse-shoe shaped zone of slightly coarser perforations (appearing opaque) on each flattened face.

Fissurina marginata (Montagu)

(Fig. 11I, J)

Characterized by its almost circular outline and thickened periphery.

Buliminella elegantissima (d'Orbigny)

(Fig. 11K, L)

Wall calcitic, consisting of radially arranged crystallites, with fine pores, glassy in appearance. Test subovoid, with chambers coiled in a high trochospiral of two to three whorls. Numerous chambers in the final whorl. Sutures slightly depressed. Aperture comma-shaped, lying within a depression. The apertural face lacks pores. Average length 0·2 mm. A marine, inner shelf species found mainly on muddy substrates. On death it may be transported into estuaries where it is deposited with mud.

(Some authors refer to this species as *B. borealis* Haynes.)

Fursenkoina fusiformis (Williamson)

(Fig. 12A–C)

Wall calcitic, of radially arranged crystallites, with fine pores (~1 μm in diameter), and glassy in appearance. Test elongate, fusiform, chambers having twisted biserial arrangement. Chambers inflated, reaching a maximum size in the mid-part of the test. Aperture terminal, comma-shaped and with a toothplate. Average length 0·3 mm. A marine species commonly transported into estuaries and deposited with muddy sediment.

This species has commonly been recorded as *Virgulina fusiformis* and sometimes as *Stainforthia fusiformis.*

Brizalina pseudopunctata (Höglund)

(Fig. 12D)

Wall calcitic, of radially arranged crystallites, perforate and glassy in appearance. Test biserial, only slightly compressed. Wall pores are present only on one half of each chamber. Sutures depressed. Aperture an arch with a toothplate, on the inner margin of the last chamber. Average length 0·3 mm. A marine shelf species commonly transported into the muddy parts of estuaries.

Brizalina variabilis (Williamson)

(Fig. 12E)

Wall calcitic, of radially arranged crystallites, perforate, often translucent in appearance. Test biserial, compressed. Chamber walls with coarse pores, each pore being set in a conical depression. Sutures depressed, straight and oblique to the margin. Aperture terminal, with a toothplate. Average length 0·5 mm. A marine inner shelf species sometimes transported into the muddy parts of estuaries.

Bolivina pseudoplicata Heron-Allen and Earland

(Fig. 12F)

Wall calcitic, of radially arranged crystallites, with moderately large pores set in depressions. Test biserial, compressed. Sutures depressed, oblique to the margin, somewhat obscured by fold-like ridges which pass along the length of the test. Aperture terminal, with toothplate. Average length 0·4 mm. An inner shelf species commonly transported into estuaries.

Bulimina marginata d'Orbigny

(Fig. 12G)

Wall calcitic, of radially arranged crystallites, very finely perforate. Test triserial, rapidly expanding, high spired. Chambers inflated, lower edge angular with small spines. Aperture comma-shaped. Average length 0·4 mm. An inner shelf, marine species.

Trifarina angulosa (Williamson)

(Fig. 12H, I)

Wall calcitic, of radially arranged crystallites, with fine pores, and glassy in appearance. Test initially triserial but the chambers are then added in a single series. Cross section is triangular. Ornament of longitudinal ribs. Aperture terminal, with a toothplate and a surrounding lip. Average length 0·4 mm. A marine shelf species sometimes transported into estuaries.

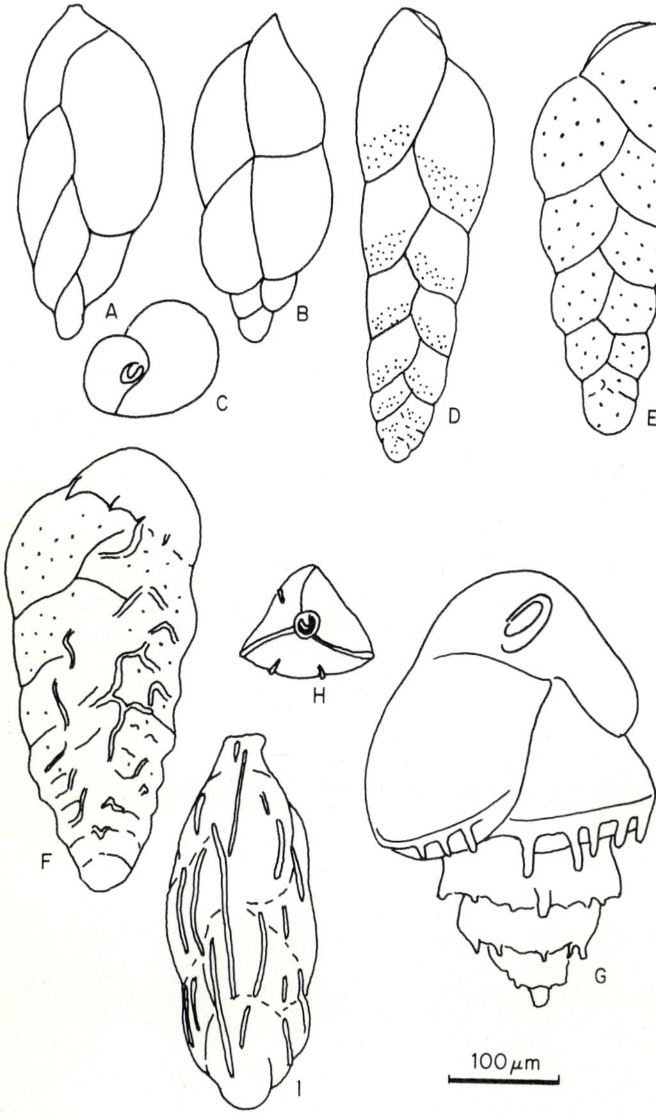

FIG. 12. Suborder Rotaliina—wall hyaline, perforate. A, B, C, *Fursenkoina fusiformis* (Williamson). D, *Brizalina pseudopunctata* (Höglund). E, *Brizalina variabilis* (Williamson). F, *Bolivina pseudoplicata* Heron-Allen and Earland. G, *Bulimina marginata* d'Orbigny. H, I, *Trifarina angulosa* (Williamson).

Elphidium

This genus is commonly known to zoologists by the junior synonym *Polystomella.* In its widest sense *Elphidium* includes all planispirally coiled foraminiferids with arch-like retral processes bridging the sutures between chambers.

Species of this genus are typical of estuaries, brackish lagoons and inner shelf seas. They show considerable variation in form and there is some overlap in morphology between species. In the past there has been great confusion in the identification of forms and so far the use of modern microscopical techniques (especially the scanning electron microscope) has not resolved the problem.

Table 1 summarizes the main morphological features of the British species. The key attempts to define the species as unequivocally as possible. Additional comments are made under the heading "Further notes on the species of *Elphidium*".

TABLE 1. Summary of the morphological features of species of *Elphidium*.

Species	Periphery	Average chambers last whorl	Evolute or involute	Aperture	Retral processes	Ornament/remarks	Average diameter (mm)
crispum	acute, keeled	18–22	involute	pores	many, long	tubercles throughout, umbilical boss	0·8
cuvillieri	round	8–10	involute	pores	6–7, short	absence of tubercles, umbilici glassy, thickened, with a few large pores	0·4
earlandi	round	8–9	involute	pores	4–6, long	tubercles in depressed areas	0·2
excavatum	round	8–9	involute	pores	3–5, very short	umbilici with calcite bosses and tubercles	0·3
gerthi	subrounded	9–11	slightly involute	pores	many, short	smooth umbilici; wall granular	0·3
incertum	round	8–11	involute	pores	few, very short	tubercles in sutures and around aperture; wall granular	0·4
macellum	acute, slight keel	17–18	involute	pores	many, long	no definite umbilical boss	0·7
magellanicum	subrounded	6–7	slightly evolute	pores	few, very short	tubercles in umbilici and along broadly depressed sutures	0·3
margaritaceum	acute	8–12	involute	pores	many, medium	covered with tiny tubercles	0·4
oceanensis	round	8–10	involute	pores	few, very short	umbilici with calcite bosses; very coarse wall pores; usually yellowish brown	0·2
williamsoni	round	10–12	involute	pores	many, long	umbilical tubercles	0·4

Key to species of *Elphidium*

1. Periphery
- Acute (see Fig. 13B) **2**
- Subrounded (see Fig. 14D) **4**
- Round (see Fig. 15C) **5**

2. Keel
- Present (see Fig. 13B) **3**
- Absent (see Fig. 13F) .... *margaritaceum* (p. 48)

3. Distinct umbilical bosses
- Present .... *crispum* (p. 48)
- Absent .... *macellum* (p. 48)

4. Umbilici
- Smooth .... *gerthi* (p. 48)
- Covered with tubercles .... *magellanicum* (p. 48)

5. Retral processes
- Long (see Fig. 16C) **6**
- Short (see Fig. 15E) **7**

6. Number of retral processes per suture
- Generally five or less .... *earlandi* (p. 50)
- Generally five or more .... *williamsoni* (p. 52)

7. Umbilical ornament
- Without boss (see Fig. 15E) **8**
- With bosses (see Fig. 15D) **9**

8. Umbilici
- Depressed .... *incertum* (p. 50)
- Raised, with thickened test, smooth .... *cuvillieri* (p. 50)

9. Pores
- Very fine .... *excavatum* (p. 50)
- Coarse .... *oceanensis* (p. 52)

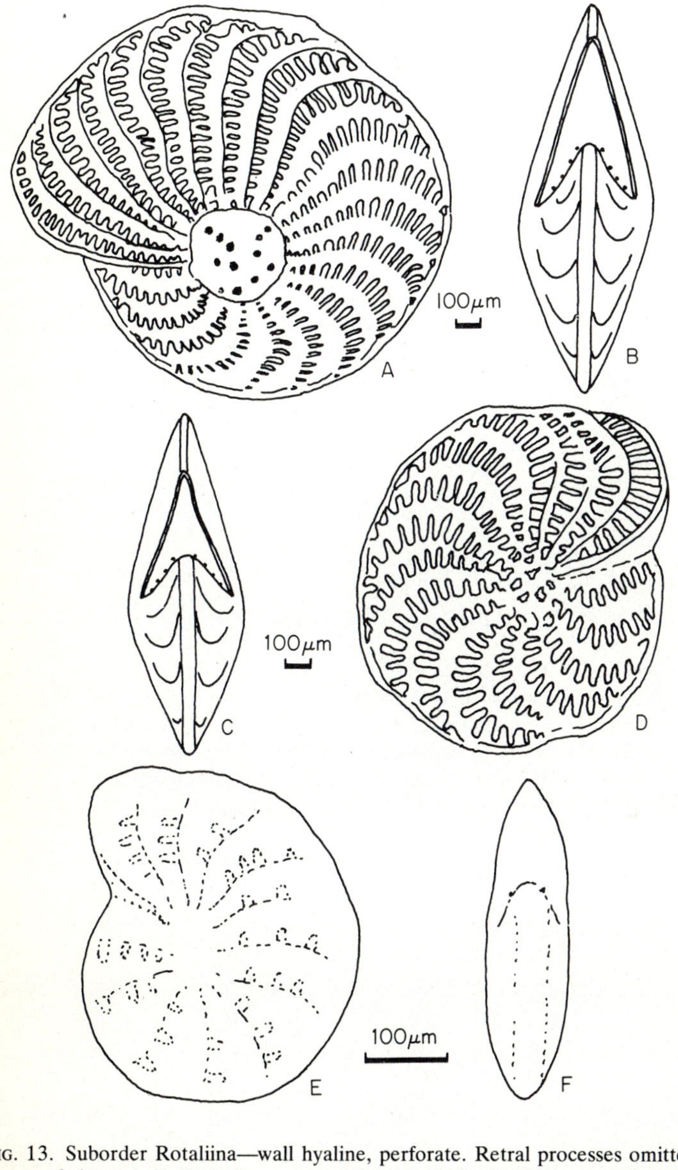

FIG. 13. Suborder Rotaliina—wall hyaline, perforate. Retral processes omitted from all apertural views. A, B, *Elphidium crispum* (Linné). C, D, *Elphidium macellum* (Fichtel and Moll). E, F, *Elphidium margaritaceum* (Cushman)—the complete test ornamentation of fine tubercles has been omitted except to outline the sutures and retral processes.

Further Notes on the Species of *Elphidium*

Periphery acute

Elphidium crispum (Linné) (Fig. 13A, B)

This is a large species characterized by massive calcite bosses in the umbilici. In large specimens these fuse to form a large boss with coarse pores.

A marine, inner shelf species of southern origin and living close to its northern limit of distribution.

Elphidium macellum (Fichtel and Moll) (Fig. 13C, D)

This large species closely resembles *E. crispum* but lacks the umbilical bosses. A complete morphological gradation exists from *macellum* without umbilical bosses via forms with small umbilical bosses to *crispum* with well developed bosses. It is doubtful whether the two species are really distinct.

The ecology is the same as for *E. crispum.*

Elphidium margaritaceum (Cushman) (Fig. 13E, F)

Readily distinguished from the preceding species by the absence of a peripheral keel and by the complete cover of tubercular ornament. A marine, inner shelf species.

Periphery subrounded

Elphidium gerthi Van Voorthuysen (Fig. 14A, B)

The subrounded periphery is due to the compression of the test. The wall structure is granular. It superficially resembles *E. earlandi* and *E. williamsoni* but both of these species have somewhat longer retral processes and they lack the smooth open umbilici of *E. gerthi.* A marine inner shelf species.

Elphidium magellanicum Heron-Allen and Earland (Fig. 14C, D)

This is an easily identified species because of the distinctive pattern of tubercular ornament in the broadly depressed sutures. A brackish to marine species.

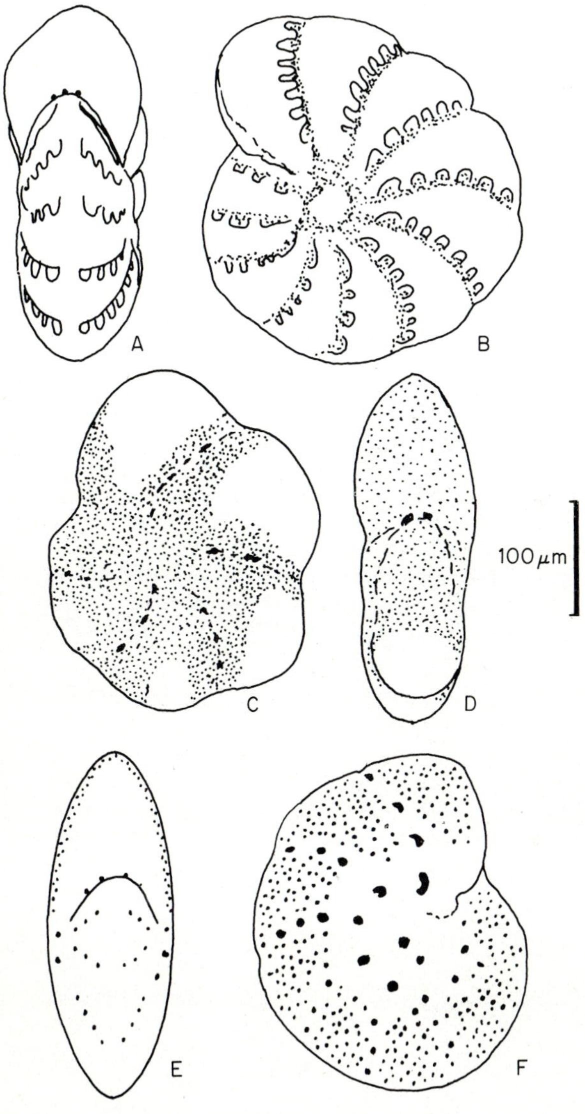

FIG. 14. Suborder Rotaliina—wall hyaline, perforate. A, B, *Elphidium gerthi* Van Voorthuysen (black dots indicate tubercles; wall pores too small to be seen). C, D, *Elphidium magellanicum* Heron-Allen and Earland (ornament as in *E. gerthi*). E, F, *Elphidium cuvillieri* Lévy (small dots are wall pores; larger dots are sutural depressions between retral processes and large openings in the thickened umbilici).

Periphery round

Elphidium cuvillieri Lévy (Fig. 14E, F)

The most distinctive features of this species are the lack of any tubercular ornament and the development of thickened, glassy umbilici. A marine, inner shelf species.

Elphidium earlandi Cushman (Fig. 15A, B)

The diagnostic features of this species are the long retral processes with only four or five (rarely six) present across each suture on each side. It is readily distinguished from *E. williamsoni*, which it superficially resembles, by the larger number of retral processes and chambers in that species. A marine, inner shelf species, sometimes found in the mouth of estuaries.

Elphidium excavatum (Terquem) (Fig. 15C, D)

This species has three to five very short retral processes on each chamber. It is a variable species and individual variants have been given separate species names (e.g. *E. clavatum, E. lidoensis, E. selseyensis*) which has caused taxonomic confusion. An inner shelf species, tolerant of brackish conditions.

Elphidium incertum (Williamson) (Fig. 15E, F)

The original illustration shows a star-shaped array of depressed sutures centred on the umbilici. A marine, inner shelf species.

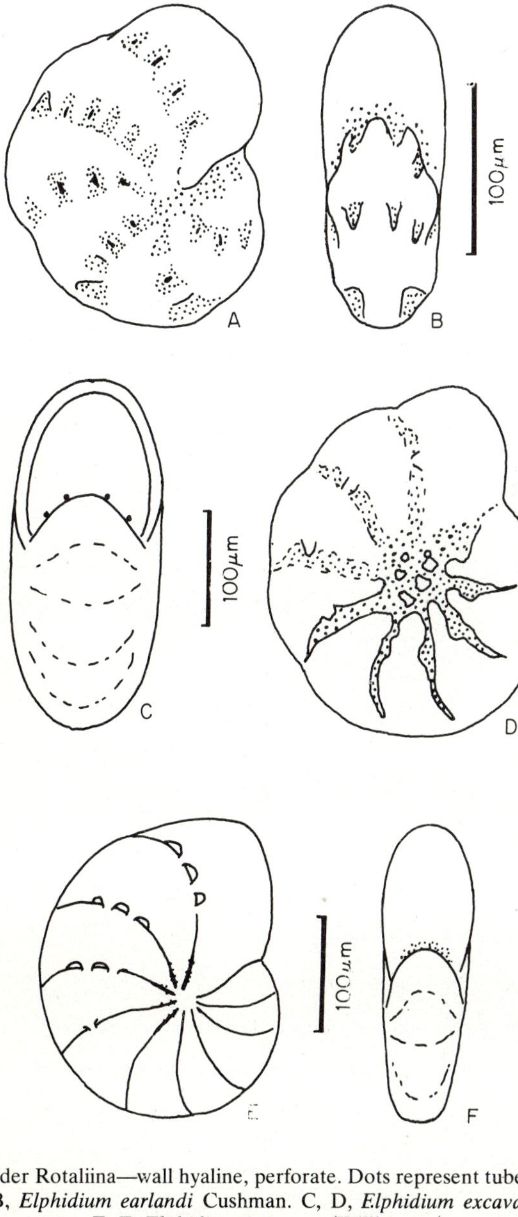

FIG. 15. Suborder Rotaliina—wall hyaline, perforate. Dots represent tubercles; wall pores omitted. A, B, *Elphidium earlandi* Cushman. C, D, *Elphidium excavatum* (Terquem). E, F, *Elphidium incertum* (Williamson).

Elphidium oceanensis (d'Orbigny) (Fig. 16A, B)

This species is readily identified because it has small retral processes, very coarse wall pores, and, unlike most other species of *Elphidium*, a yellowish brown coloured wall. It has been recorded under the names *E. gunteri* Cole and *E. waddensis* (Van Voorthuysen). A brackish to marine species.

Elphidium williamsoni Haynes (Fig. 16C, D)

The diagnostic features are the numerous, long, retral processes and the greater than average number of chambers in the outer whorl (10–12). It is sometimes confused with *E. earlandi* but the latter has fewer retral processes and generally only eight to nine chambers in the outer whorl.

In the literature this species has been misidentified as *E. excavatum* and *E. articulatum* (d'Orbigny).

This is often the commonest brackish species of *Elphidium* as it has a very wide tolerance range for salinity variation.

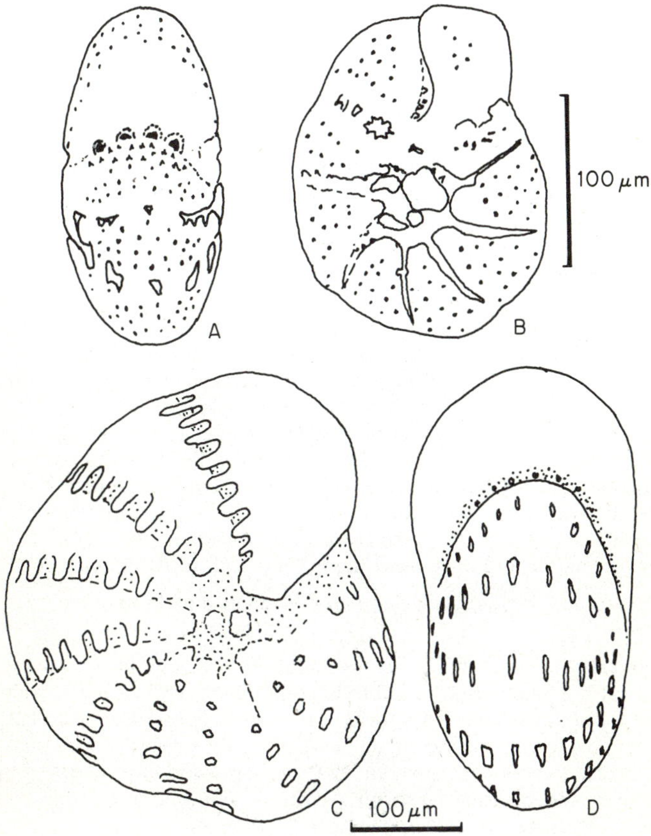

FIG. 16. Suborder Rotaliina—wall hyaline, perforate. A, B, *Elphidium oceanensis* (d'Orbigny) (dots represent coarse wall pores). C, D, *Elphidium williamsoni* Haynes (dots represent small tubercles).

Protelphidium germanicum (Ehrenberg)
or
Nonion germanicum (Ehrenberg)
(Fig. 17A, B)

Wall calcitic, of radially arranged crystallites, finely perforate, glassy in appearance in unetched specimens. Test planispiral with five to eleven chambers in the last whorl. Sutures slightly depressed over the periphery but deeply depressed close to the umbilici. Tubercular ornament extends from the umbilici into the inner depressed sutures. The aperture is a row of pores at the base of the last chamber. Average greatest diameter 0·4 mm. This is an extremely euryhaline species with a salinity tolerance of 0–35‰. It is abundant in brackish water in estuaries and lagoons.

The naming of the species has been the subject of much confusion and controversy. It has been misidentified as *Nonion depressulus* from which it is readily separated on wall structure and on morphology. It has been independently named the following species: *Nonion tisburyensis* Butcher, 1948, *Nonion paralium* Tintant, 1954, and *Protelphidium anglicum* Murray, 1965. It now seems beyond doubt that the correct species name is *germanicum*. There is then the problem of genus. Basically the difficulty is the value of wall structure in classification. According to common practice the genus *Nonion* has a granular wall while that of *Protelphidium* is radial. Hansen and Anderson (1976) prefer to use this character to distinguish species rather than genera. This is a subjective matter and at present there is no uniformity of practice.

Nonion depressulus (Walker and Jacob)
(Fig. 17C, D)

Wall calcitic, of granular arrangement, translucent in appearance. Test a somewhat flared planispiral, laterally compressed, with seven to eleven chambers in the outer whorl. Sutures flush to slightly depressed in the early portion but later strongly depressed. Umbilici depressed or raised if wall thickening has taken place. Tubercular ornament fills the depressed inner parts of the sutures and covers the umbilici. Aperture a series of pores at the base of the last chamber. Average greatest diameter 0·3 mm.

An inner shelf, essentially marine species which is found in estuary mouths. In the older literature *N. depressulus* often includes forms now called *Protelphidium germanicum.*

Buccella frigida (Cushman)
(Fig. 17E–G)

Wall calcitic, of radially arranged crystallites, with extremely fine pores, often translucent in appearance. Test biconvex, almost circular in plan, chambers coiled in a trochospiral. Sutures on the spiral side flush; those of the umbilical side broadly depressed and filled with tubercular ornament which also extends into the umbilicus. Within the ornament there are hidden additional apertural openings on the sutures. The main aperture at the base of the final chamber is a slit usually obscured by the tubercular ornament. Average greatest diameter 0·2 mm. A cool water, mainly marine, species which can tolerate slightly brackish conditions. The English Channel represents the southern limit of distribution.

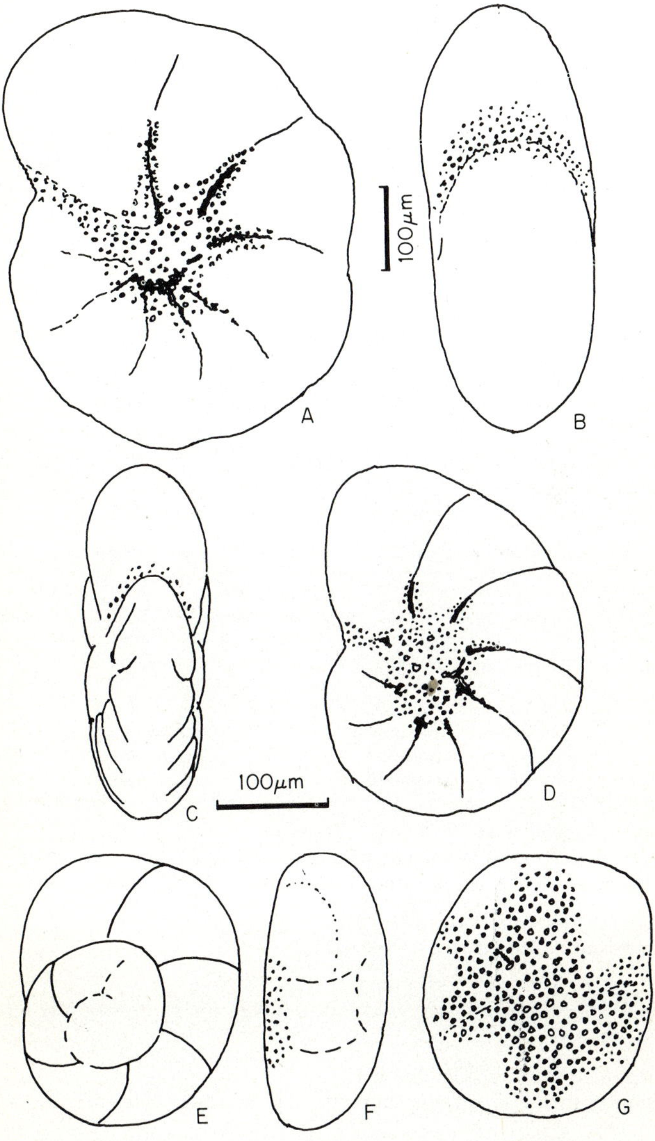

FIG. 17. Suborder Rotaliina—wall hyaline, perforate. Dots and circles represent tubercles; wall pores omitted. A, B, *Protelphidium germanicum* (Ehrenberg). C, D, *Nonion depressulus* (Walker and Jacob). E, F, G, *Buccella frigida* (Cushman).

Ammonia

The genus *Ammonia* is characteristic of marginal marine environments, i.e. lagoons, estuaries and inner shelf. The representatives are variable in form. Two different attitudes exist. Some authors split the variants into species, but, because there is a morphological overlap, recognition is subjective. Other authors believe that only one species is present and that the variants are produced by the influence of the environment. There is some experimental support for this latter view (Schnitker, 1974). Here the British variants are referred to *Ammonia beccarii* (Linné). However, details are given below of the features which characterize the other "species" (here regarded as variants).

Ammonia beccarii (Linné)
(Fig. 18A–L)

Wall calcitic, of radially arranged crystallites, with pores, glassy or translucent, and sometimes brownish in colour. Test biconvex, subcircular in outline. Chambers trochospirally coiled, six to nine in the last whorl. Sutures on spiral side flush to depressed; those on the umbilical side depressed. Umbilicus open or occupied by a calcite boss. Aperture a slit at the base of the last chamber, extending into the umbilicus beneath a flap. Average greatest diameter 0·4 mm.

TABLE 2. Morphological details of variants of *Ammonia beccarii.*

Species	Spiral side	Periphery	Number of chambers visible on umbilical side	Umbilical ornament	Average greatest diameter (mm)
aberdoveyensis	low, conical	rounded	8–9	none	0·4
batavus	flattened	subangular	8–9	calcite plug	0·4
limnetes	flattened	rounded	6–7	none	0·3
tepida	low conical	rounded	6–8	none	0·4

Variant *aberdoveyensis* Haynes (Fig. 18A–C)

Differs from *batavus* in lacking an umbilical boss and from *tepida* in its growth rate, there being eight to nine rather than six to eight chambers per whorl.

Variant *batavus* Hofker (Fig. 18D–F)

Characterized by its umbilical boss.

Variant *limnetes* Todd and Brönnimann (Fig. 18G–I)

The small size, flat spiral side, and inflections in the sutures on the umbilical side are distinctive.

Variant *tepida* Cushman (Fig. 18J–L)

Very similar to *aberdoveyensis* but with slightly fewer chambers per whorl.

In Britain variants without umbilical bosses are common in estuaries and brackish lagoons. They can tolerate a diurnal salinity variation of 0–35‰. The optimum temperature range is 15–20°C. Variants with an umbilical boss (*batavus*) are found in more marine, inner shelf environments.

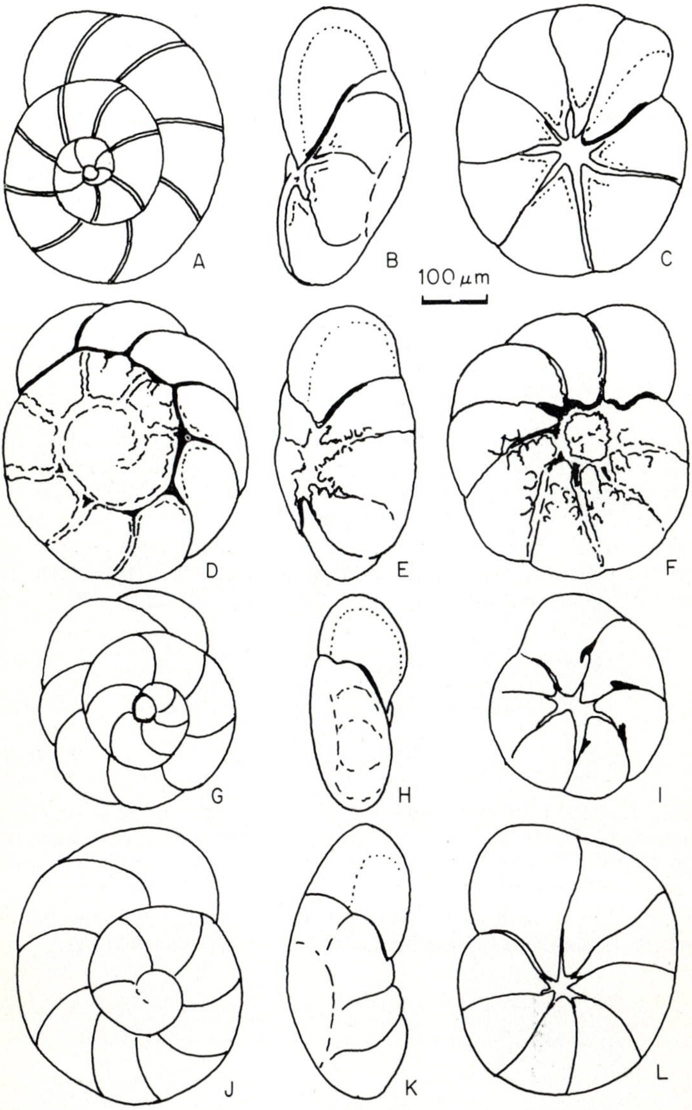

FIG. 18. Suborder Rotaliina—wall hyaline, perforate. The pores are very fine and have been omitted. The dots represent faint changes in wall texture and delimit the area of the apertural face which is imperforate. *Ammonia beccarii* (Linné) variants: A, B, C, *aberdoveyensis* Haynes; D, E, F, *batavus* Hofker; G, H, I, *limnetes* Todd and Brönnimann; J, K, L, *tepida* Cushman.

Gavelinopsis praegeri (Heron-Allen and Earland)

(Fig. 19A–C)

Wall calcitic, of radially arranged crystallites. Test trochospirally coiled, planoconvex, with five to six chambers in the last whorl. Sutures on spiral side flush and strongly curved backwards, on umbilical side depressed and radial. Umbilicus with large flat boss. Aperture at base of final chamber on umbilical side. Average greatest diameter 0·2 mm. An inner shelf, marine species, which lives clinging to firm substrates.

Rosalina anomala Terquem

(Fig. 19D–F)

Wall calcitic, coarsely perforate on spiral side, imperforate on umbilical side. Test trochospirally coiled, five to six chambers in final whorl. Sutures on spiral side slightly depressed and curved backwards, on umbilical side depressed and radial. Umbilicus open. Primary aperture a slit at base of final chamber. Secondary apertures open into umbilicus from chamber flaps. Average greatest diameter 0·5 mm. An inner shelf, marine species, which lives clinging to seaweeds and other firm substrates, with the umbilical side as the surface of attachment.

Rosalina williamsoni (Chapman and Parr)

(Fig. 19G–I)

Wall calcitic, finely perforate and very thin. Test trochospirally coiled, with low convex spiral side and flat to concave umbilical side. Periphery acute. Normally six chambers visible in the last whorl. Sutures on spiral side strongly curved backwards, those on umbilical side curved to radial. Umbilicus sometimes occupied by a small boss. Average greatest diameter 0·3 mm. An inner shelf, marine species which probably lives clinging to firm substrates.

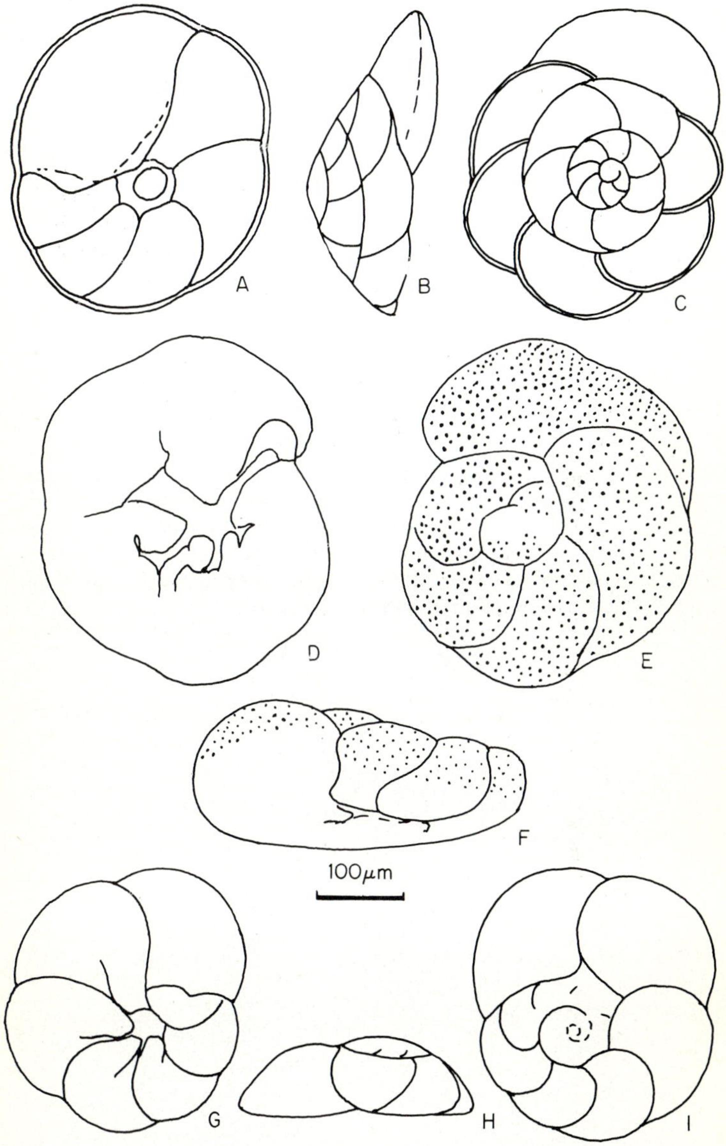

FIG. 19. Suborder Rotaliina—wall hyaline, perforate. A, B, C, *Gavelinopsis praegeri* (Heron-Allen and Earland). D, E, F, *Rosalina anomala* Terquem. G, H, I, *Rosalina williamsoni* (Chapman and Parr).

Glabratella millettii (Wright)

(Fig. 20A–C)

Wall calcitic, finely perforate. Test trochospirally coiled, with low convex spiral side and slightly concave umbilical side. Sutures on spiral side flush, strongly curved backwards; those on umbilical side slightly depressed and radial to sigmoid, obscured by radiating tubercular ornament. Periphery acute. Aperture an arch at the base of the final chamber with a second opening into the umbilicus beneath a chamber flap. Average greatest diameter 0·3 mm. An inner shelf, marine, species which probably lives clinging to firm substrates.

Asterigerinata mamilla (Williamson)

(Fig. 20D–F)

Wall calcitic, finely perforate throughout but with large pores close to the sutures. Test trochospirally coiled, with convex spiral side and planar umbilical side. Sutures on spiral side thickened and curved backwards; those on umbilical side only slightly depressed. Umbilicus closed and surrounded by secondary chamberlets. Aperture an arch at the base of the last chamber. Average greatest diameter 0·3 mm. An inner shelf, marine, species.

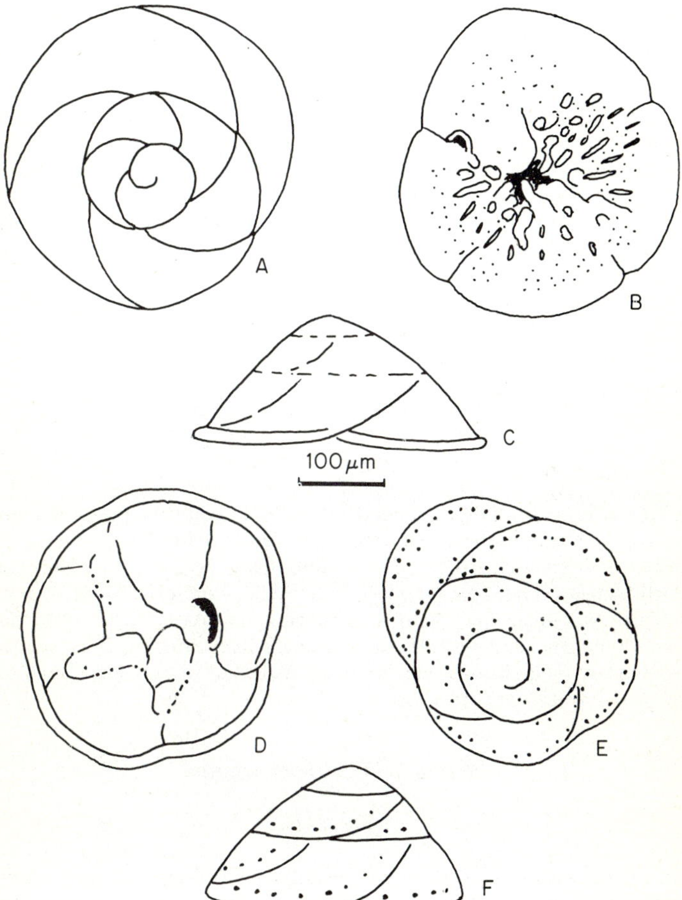

FIG. 20. Suborder Rotaliina—wall hyaline, perforate. A, B, C, *Glabratella millettii* (Wright). D, E, F, *Asterigerinata mamilla* (Williamson).

Cibicides lobatulus (Walker and Jacob)

(Fig. 21A–C)

Wall calcitic, of radial and granular nature, with large pores, often translucent to opaque in appearance. Test variable in shape, commonly planoconvex, trochospirally coiled, with a plane spiral side and convex umbilical side. Periphery angular and with an imperforate thickening of the wall. Sutures depressed. Aperture an arch and slit, bordered by a lip and extending along the spiral suture on the spiral side. Average greatest diameter 0·8 mm. A marine, inner shelf species, which clings to firm substrates such as seaweeds, tunicates, shells and rocks, in areas subject to disturbance by currents. Dead tests are transported into estuary mouths.

Planorbulina mediterranensis d'Orbigny

(Fig. 21D, E)

Wall calcitic, of radially arranged crystallites, coarsely perforate, glassy in appearance and sometimes faintly brown in colour. Test planoconvex, quadrate, polygonal or somewhat irregular in outline. Chambers trochospirally coiled in the early stages, later being added in a cyclical pattern. Sutures depressed. Numerous apertures, each with a bordering lip. Average greatest diameter 0·5 mm. A marine, inner shelf species which lives clinging to firm substrates such as seaweeds, shells or rocks, in areas disturbed by currents. Dead tests are transported into estuary mouths.

Acervulina inhaerens Schultze

(Fig. 21F)

Wall calcitic, of radially arranged crystallites with large pores in depressions, generally white and opaque in appearance. Test irregular in form, usually being controlled by the topography of the substrate which the individual is encrusting. Chambers inflated, initially coiled trochospirally but later becoming added in an irregular pattern. Numerous apertures. Average length 0·6 mm. A marine, inner shelf species which encrusts firm substrates. On death it may be swept into estuary mouths.

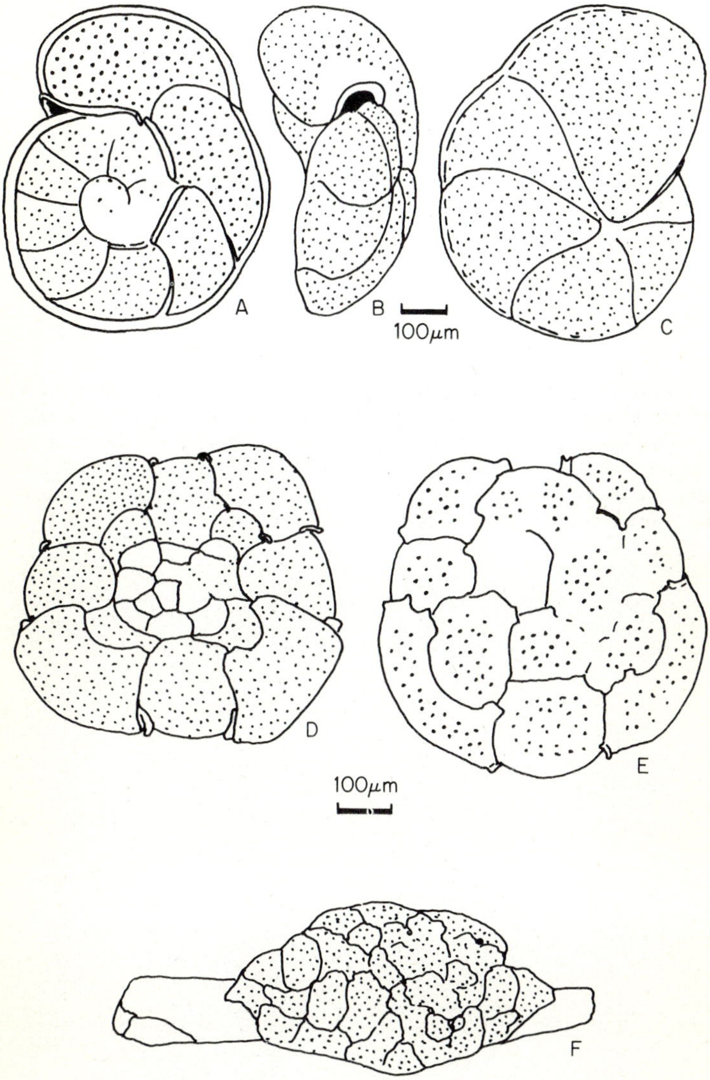

FIG. 21. Suborder Rotaliina—wall hyaline, perforate. Dots represent wall pores. A, B, C, *Cibicides lobatulus* (Walker and Jacob). D, E, *Planorbulina mediterranensis* d'Orbigny. F, *Acervulina inhaerens* Schultze, attached to a shell fragment.

Cassidulina obtusa Williamson

(Fig. 22A–C)

Wall calcitic, granular. Test biconvex, coiled on an enrolled biserial plan. Sutures depressed. Aperture a slit with bordering lip in a depression at the base of last chamber. Average greatest diameter 0·2 mm. A shelf species living on muddy substrates and sometimes transported onto intertidal mudflats.

Lamarckina haliotidea (Heron-Allen and Earland)

(Fig. 22D–F)

Wall aragonitic, extremely finely perforate, very thin. Test coiled on a trochospiral plan. Spiral side convex, umbilical side flat. Periphery angular, with a keel. Sutures depressed throughout. Umbilicus broad and deep and shows early part of whorls. Aperture opens into umbilicus. Average greatest diameter 0·2 mm. An inner shelf, marine, species which lives on muddy substrates and is readily transported because of its small size and light test.

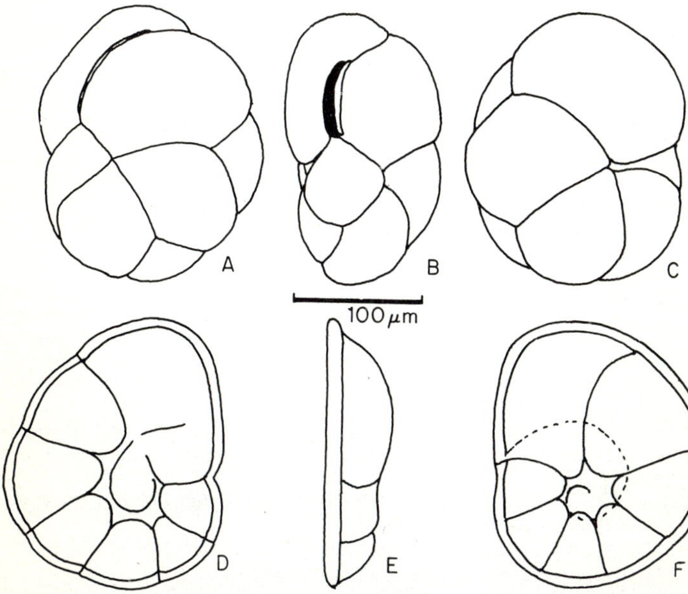

FIG. 22. Suborder Rotaliina—wall hyaline, perforate. A, B, C, *Cassidulina obtusa* Williamson. D, E, F, *Lamarckina haliotidea* (Heron-Allen and Earland).

Acknowledgements

I wish to thank Mrs G. C. Wright for typing the manuscript and Dr J. R. Haynes (Aberystwyth) for offering helpful comment. Professor D. Nichols kindly allowed access to a phase contrast microscope. The frontispiece is the work of Mr J. Jones.

References

HANSEN, H. J. and ANDERSON, A. L. L. 1976. Wall structure and classification of fossil and recent elphidiid and nonionid Foraminifera. *Fossils and Strata* **10,** 1–37.

LOEBLICH, A. R., Jr., and TAPPAN, H. 1964. Protista 2, Sarcodina, chiefly "thecameobians" and Foraminiferida. In *Treatise on Invertebrate Paleontology* (Ed. R. C. Moore), Part C, Vol. 1. pp. C1–510; Vol. 2. pp. C511–900. University of Kansas Press and Geological Society of America.

LOEBLICH, A. R., Jr., and TAPPAN, H. 1974. Recent advances in the classification of the Foraminiferida. In *Foraminifera* (Eds R. H. Hedley and C. G. Adams), Vol. 1. pp. 1–53. Academic Press, London and New York.

MURRAY, J. W. 1971. *An Atlas of British Recent Foraminiferids*, xii + 244 pp. Heinemann Educational Books, London.

MURRAY, J. W. 1973. *Distribution and Ecology of Living Benthic Foraminiferids*, xii – 274 pp. Heinemann Educational Books, London.

SCHNITKER, D. 1974. Ecotypic variation in *Ammonia beccarii* (Linné). *J. foramin. Res.* **4,** 216–223.

Further Reading

ARNOLD, Z. 1974. Field and laboratory techniques for the study of living Foraminifera. In *Foraminifera* (Eds R. H. Hedley and C. G. Adams). Vol. 1. pp. 153–206. Academic Press, London and New York.

HAYNES, J. R. 1973. Cardigan Bay recent Foraminifera. *Bull. Br. Mus. nat. Hist. Zool.* Suppl. 4, 1–245.

HEDLEY, R. H. 1964. The biology of Foraminifera. *Int. Rev. gen. exp. Zool.* **1,** 1–45.

Index of Genera and Species

Notes

Notes